BUILDING A STRAW BALE HOUSE

The Red Feather Construction Handbook

Published by
Princeton Architectural Press
37 East Seventh Street
New York, New York 10003

For a free catalog of books, call 1.800.722.6657.
Visit our web site at www.papress.com.

Printed and bound in China
09 08 07 06 05 5 4 3 2 1 First edition

Editing: Linda Lee and Megan Carey
Design: Michael Lindsay, studiovertex

Library of Congress Cataloging-in-Publication Data
Corum, Nathaniel.
Building a straw bale house : the Red Feather construction handbook / Nathaniel Corum.—1st ed.
p. cm.
Includes bibliographical references.
ISBN 1-56898-514-2 (pbk. : alk. paper)
1. Straw bale houses—Design and construction—Handbooks, manuals, etc. 2. Indians of North America—Housing—South Dakota. 3. Red Feather Development Group. I. Title.

TH4818.S77C67 2005
693'.997—dc22

2005001751

IMAGE CREDITS

Wayne Bastrup: 118, 119
Skip Baumhower I www.baumhower.com: cover, xv, xvi, xxii, xxvi, 9, 12, 13, 15 top, 17, 20, 22, 23, 25, 29 left, 33, 36, 37, 39 bottom, 40, 41, 43, 45 bottom left, 45 bottom right, 46, 47 bottom left, 47 bottom right, 48, 49, 52, 55, 56 center right, 61 top right, 62, 63 bottom, 67, 69, 70 bottom, 74–77, 80, 81, 83, 87 top, 90–93, 105 top, 112, 114, 116, 117, 130, 134, 139, 146, 148, 150, 156 top, 160 bottom, 162, 165, 166, 172, 173, 175, 178
Richard K. Begay Jr.: 128
Richard K. Begay Jr., Corbin Plays, and Nathaniel Corum: 126, 127, 129
Jonathan Corum: 170, 171
Nathaniel Corum: 6, 7, 98, 121, 122, 123, 140, 145, 155 bottom, 159 bottom, 160 top
Nathaniel Corum and Corbin Plays: vii, 3, 11, 15 bottom, 19, 27, 35, 51, 65, 79, 95, 101, 124, 125
David S. Holloway/Apix: xii
Matts Myhrman: viii, xxiv, xxv
Michael Rosenberg: x, xviii, xx, 4, 5, 28, 29 right, 30, 31, 39 top, 45 top, 47 top, 56 top row, 56 center left, 56 bottom row, 59, 61 top, 61 bottom right, 66, 70 top, 73, 87 bottom, 88, 96, 97, 99, 105 bottom row, 106, 107, 109, 110, 152, 155 top, 156 bottom, 159 top, 167, 168, 174, 176, 177
Jim Waters: 63 top, 102

BUILDING A STRAW BALE HOUSE

The Red Feather Construction Handbook

NATHANIEL CORUM

Foreword by Dr. Jane Goodall

Princeton Architectural Press, New York

This book is not a substitute for technical design and construction documents prepared by qualified professionals. It is not intended to be a definitive "how-to" book for a particular construction project. It provides an overview of some of the processes that Red Feather Development Group has developed in the course of managing a variety of past projects, no two being exactly the same. Just as our experience has varied from project to project, please expect to encounter situations that are not covered in this book. Every construction project should involve qualified design and construction professionals in accordance with applicable building and safety codes and authorities.

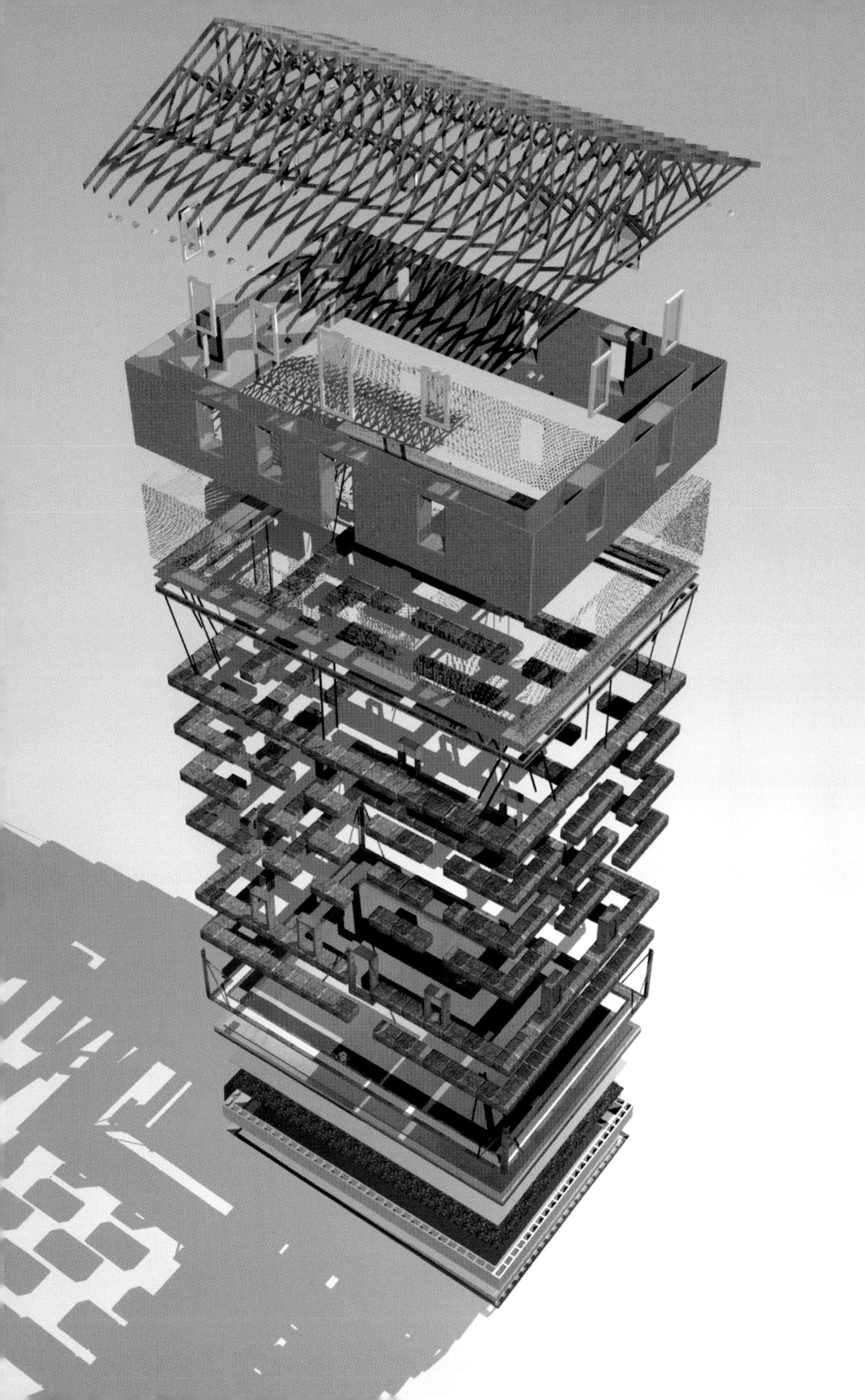

The Simonton Family Home, Purdum, Nebraska, built c. 1908: An early example of a load-bearing bale building that is said to have been constructed with prairie hay cut after a frost.

And what is it to work with love?
...It is to build a house with affection,
even as if your beloved were to dwell in that house.

Kahlil Gibran

CONTENTS

[xiii] **Foreword**

[xvii] **Mission Statement**

[xix] **Preface**

[xxiii] **Introduction**

HANDBOOK

[2] Chapter 1: **Foundation and Site Work**

[10] Chapter 2: **Roof Bearing Assembly**

[18] Chapter 3: **Base Plates**

[26] Chapter 4: **Window and Door Bucks**

[34] Chapter 5: **The Straw Bale Wall**

[50] Chapter 6: **Above the Straw Bale Wall**

[64] Chapter 7: **Lath**

[78] Chapter 8: **Stucco**

[94] Chapter 9: **Interior Walls**

[100] Chapter 10: **Finish Details**

APPENDICES

[113] Appendix A: **Safety**

[117] Appendix B: **Design Drawings**

[131] Appendix C: **Time and Labor**

[135] Appendix D: **Tools and Materials**

[141] Appendix E: **Radiant Floor System**

[147] Appendix F: **Straw Bale References**

CASE STUDIES

[153] **Red Feather Projects**

[163] **Turtle Mountain: Anatomy of a Build**

[169] **Red Feather Development Group**

[179] **Contributors**

[181] **Contact Information**

Dr. Jane Goodall, DBE, U.N. Messenger of Peace

FOREWORD | DR. JANE GOODALL

A story of compassion, determination, and courage lies behind the Red Feather project. A decade ago, Robert Young read how three Native American elders froze to death during the winter in South Dakota due to a severe lack of proper housing. He then traveled to the Pine Ridge Reservation and was shocked by conditions on the Indian reservation—poverty, overcrowding, hunger, and hopelessness.

Back home in Seattle, Robert sought solutions. With a group of friends and family he went back to South Dakota and gathered the materials for the construction of the first house—a home for Oglala Lakota elder Katherine Red Feather. This initial project was successful and, upon its completion, Robert asked Katherine if he could establish a non-profit housing organization in her honor.

Red Feather Development Group was born, and more houses were built with the help of reservation members. Other environmentally and culturally sustainable projects were designed to increase self esteem within the communities. Now Red Feather has grown in scope and purpose. In 2003 Red Feather moved to Montana in order to be closer to their community partners and has added full-time directors of development, program, and community design. Forty housing-based projects have been completed to date and more projects are now on the drawing board.

Through experience, the Red Feather team has learned much about the provision of ecologically and socially responsible housing. This book—*Building a Straw Bale House* by Community Design Director Nathaniel Corum—shares Red Feather's vision and process. This book is a timely and important tool toward the empowerment of communities facing housing deficits.

I met Robert when he received the Volvo for Life Award at a celebration in New York. We became excited about the prospect of collaboration. The Jane Goodall Institute has developed similar solutions to problems of poverty. In Africa we work with villagers living in extreme poverty, encouraging them to become involved in environmentally and culturally sustainable projects. Our Roots & Shoots program, now in eighty-seven countries around the world, encourages youth (pre-school through university) to participate in service projects in their schools and communities. In poor rural areas and in the inner city we have seen them acquire self esteem and the respect of others.

We are working to introduce the Roots & Shoots program into Native American reservations. Red Feather can powerfully help us: Roots & Shoots can provide additional opportunities for the young people involved in Red Feather projects.

That so many people living in the twenty-first century, in the richest country in the world, should be living in abject poverty, is a blot on American society. The Red Feather project is extremely important. It is culturally and environmentally sensitive; it has incredibly low overheads; it is truly making a difference.

Jane Goodall PhD, DBE
Founder—the Jane Goodall Institute
United Nations Messenger of Peace
www.janegoodall.org

RED FEATHER DEVELOPMENT GROUP

Mission Statement

Red Feather focuses the public's eye on the poverty and dire housing problems that have plagued American Indian reservations for generations. It raises funds and organizes volunteers to build homes within reservation communities, and it empowers tribal members to create long-term, sustainable housing solutions. Its work is made possible by the concern and steadfast generosity of its many sponsors and volunteers.

PREFACE

Like a barn raising, this book is founded on the efforts of a group of people collaborating with American Indian communities to build sustainable and affordable housing. To tell the story of the American Indian Sustainable Housing Initiative, I must first thank and praise the many individuals whose efforts have made culturally appropriate straw bale housing a cost-effective reality for several tribal nations.

Since Red Feather began constructing housing some five years ago, the nonprofit has enjoyed an informative and fruitful relationship with two academic institutions—the University of Washington College of Architecture and Urban Planning (CAUP) and Pennsylvania State University—and with Professors David Riley, Sergio Palleroni, Scott Wing, and CAUP Dean Robert Mugerauer, who also encouraged my early education in sustainability at University of Texas at Austin.

The larger straw bale community—many of whose members I sought out in anticipation of my partnership with Red Feather—welcomed my inquiry and freely shared the hard-won lessons and technical developments that have graced the "straw bale revival" for the past twenty years. Matts Myhrman and Judy Knox were especially helpful by reviewing this book, attending several Red Feather builds, and sharing their considerable straw bale wisdom. Athena and Bill Steen, David Eisenberg, and Chris Magwood also shed light onto this project by bringing, through their continuing work, the current thinking on straw bale design and construction to all of us.

Without the support of Jonathan Rose, the Enterprise Foundation, and the Rose Architectural Fellowship, my work would not be possible. Thank you, Jonathan, for believing that design matters when we go to build affordable housing and for selecting Stephen Goldsmith, a bastion of advice, intelligence, and leadership, to be fellowship director. This book also benefits significantly from the involvement of my wonderful editors at the Princeton Architectural Press. Thank you Megan Carey, Linda Lee, and Clare Jacobson for your care, and excitement.

Without the vision and leadership of Robert Young, the Red Feather Development Group would not exist. Thus my deepest gratitude goes to Robert and Anita Young for getting it all started and for making work feel like home in so many ways. A debt of gratitude also goes to Ryan Batura, former construction program director (CPD), Wayne Bastrup, former construction coordinator, and Michael Kelly, current CPD, all of whom helped develop the procedures and processes for Red Feather buildings.

The professional design community has likewise played an invaluable role. I had my first experience working in straw bale with architect James Bell. More recently architects Don McLaughlin and Alfred VonBachmayr, and engineers Tom Beaudette and Art Fust provided crucial support and counsel. As the School of Architecture at Montana State University (MSU) provides an important academic affiliation

in Bozeman, I want to thank Professor Lori Ryker, the students in our MSU Straw Bale Studio, and Dean Clark Llewellyn, for his friendship and support.

Without Corbin Plays, Richard Begay, Scott Frasier, Stacie Laducer, Mark Roundstone, Mary Tenakongva, and Elmer Yazzie, I would not be in a position to understand the housing needs of American Indians. Thank you all for your patience and your friendship.

Finally I would like to thank my family: Richard, Sarah, Morgan, Jan, and Aranye. They taught me to care, build, and write—and they continue to be my inspiration. This book is dedicated to them and written in thanks to all who build shelter for those who need it.

Nathaniel Corum
Bozeman, Montana
April 13, 2005

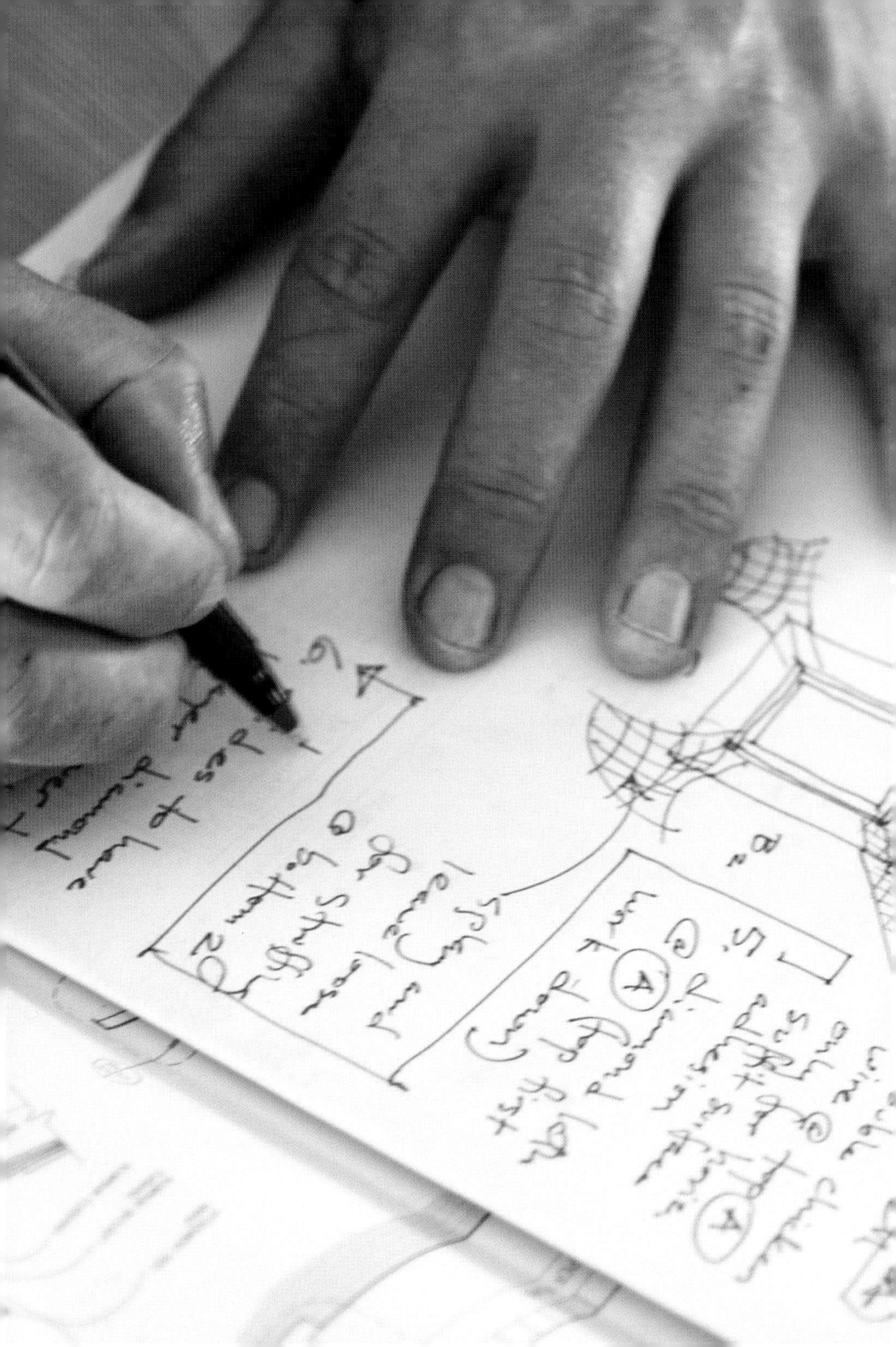

INTRODUCTION

The building detailed in the pages below is the best house the Red Feather Development Group presently knows how to build; thus there is a great deal of knowledge and experience embedded in this prototype. But since we learn something new on every building project, and as straw bale construction is in a remarkable phase of growth with new and better techniques being invented each season, at this time we can only give you a recipe for Red Feather's best building so far, mindful that we will no doubt have further improvements to offer after these pages reach the press.

However, by following the suggestions in this handbook, you will be spared many of the pitfalls of building with straw bale, having benefited significantly from our mistakes. Nevertheless, we encourage readers to consult other straw bale books on the market if they are interested in techniques and design variations beyond the scope of this handbook.

There are infinite ways to build a house. Over the past decade Red Feather has researched and tried many construction methods and has arrived at a prototype that reflects the needs and resources of the communities we serve. In writing this handbook, we have further refined the typical Red Feather house. Our criteria in making improvements is summarized in the following list:

1. Ease of Construction. We present a straightforward process that we know will work.

2. Low Cost. Our objective is to get the most house for the least expense.

3. Ease of Community Involvement. The process described here utilizes volunteer and community labor, and so we have (wherever possible) reduced the need for professional expertise.

4. Environmentally Sound. The house presented here is built of materials that, whenever possible, are local in origin, non-toxic in composition, and energy-efficient in practice.

5. Long Life/Loose Fit. We offer a building that will last, one that easily adapts to the changing needs of its inhabitants.

The Burke Homestead, near Alliance, Nebraska, built c. 1903: One of the oldest known bale buildings, the house is still standing after over one century. Despite the fact that the house has had little maintenance since the 1950s, the bale walls were still in good condition when they were sampled in 1993.

Qualifications

This handbook gives the reader information about building a straw bale home from initial site selection to finished product. However, portions of this process should be performed by professionals. For example, we assume that you will engage a licensed contractor to do the foundation, plumbing, and electrical work.

Safety is paramount, and building projects should be managed by professionals who are experienced in safety-conscious construction procedures. See Appendix A for some safety guidelines that have worked for our organization.

The Scott House, near Garden, Nebraska, built c. 1935: A horse-powered baler was used to bale local wheat for this load-bearing straw bale home. Still in great condition, the house is a testament to the longevity of straw bale buildings.

The construction process outlined in this book is based upon the Red Feather Development Group's experience on numerous projects. Some builders and their communities may wish to expand on the design of Red Feather's basic prototype. While we encourage elaboration, we also caution readers against making modifications without the assistance of a licensed architect or a design engineer.

For those new to straw bale construction

A straw house? Yes! People have been building with straw for millennia in places all over the world. From the grass huts of pre-history to the thatched roofs of Northern Europe, straw has long been a building material. About a century ago, after the advent of hay-baling machinery, farmers in tree-scarce Nebraska developed a way of building homes with baled straw. Some of these early straw bale buildings still exist and continue to perform well. Like jazz music, straw bale construction is a native, American technology. Moreover, during the last few decades, the United States has become increasingly interested in straw bale construction. As a result, much has been learned, and building methods have vastly improved, in part because these structures are intensively tested and researched, and thus prove straw bale construction to be a viable and code-compliant method of building. Straw bale buildings have been constructed in Alaska and most—if not all—of the continental United States. Except in localities with excessively humid climates, such as the Gulf Coast and the Hawaiian Islands, straw bale is an appropriate technology. Straw is a particularly good choice in cold and dry places—like Northern Plains and the American Southwest—where super-insulated buildings are desirable.

Still, skeptics abound, so we may need to explain the obvious: straw is not hay. Straw bales used for building are an agricultural by-product. Whether left-over stalks of wheat, rice, or a similar plant are used, the nutritional portion of the grain has been extracted. What is left—the stalks (or "culm," to adopt the agricultural term)—has little food value. This is good news for bale construction because insects and animals have no interest in eating straw bale buildings. After being coated with stucco, baled straw is resistant to fire. But just as poorly maintained conventional jobsites become dangerously flammable when wood scraps and sawdust are left lying about, so builders using this handbook should keep loose straw cleaned up and away from sources of combustion. Obviously welding, smoking, and other open-flame activities should be kept well away from a pre-stucco straw bale building.

The Martin/Monhart House, Arthur, Nebraska, built c. 1925: Still in fine shape after over seventy-five years, this bale house features a useable attic space consisting of two small upstairs rooms.

HANDBOOK

The Red Feather Straw Bale Process

Your home's position relative to the land on which it sits has significant effects on its beauty and thermal efficiency. Consequently, you will want to spend some time considering how to orient your house to make the best use of your building site. There are several factors to consider here: local customs, potential views, windbreaks, existing roads and infrastructure, existing utilities locations, the warming path of the sun, as well as your preferences and intuitions. If at all possible, you might want to visit the land in different seasons, and talk with neighbors who have experience placing buildings in the vicinity of your site.

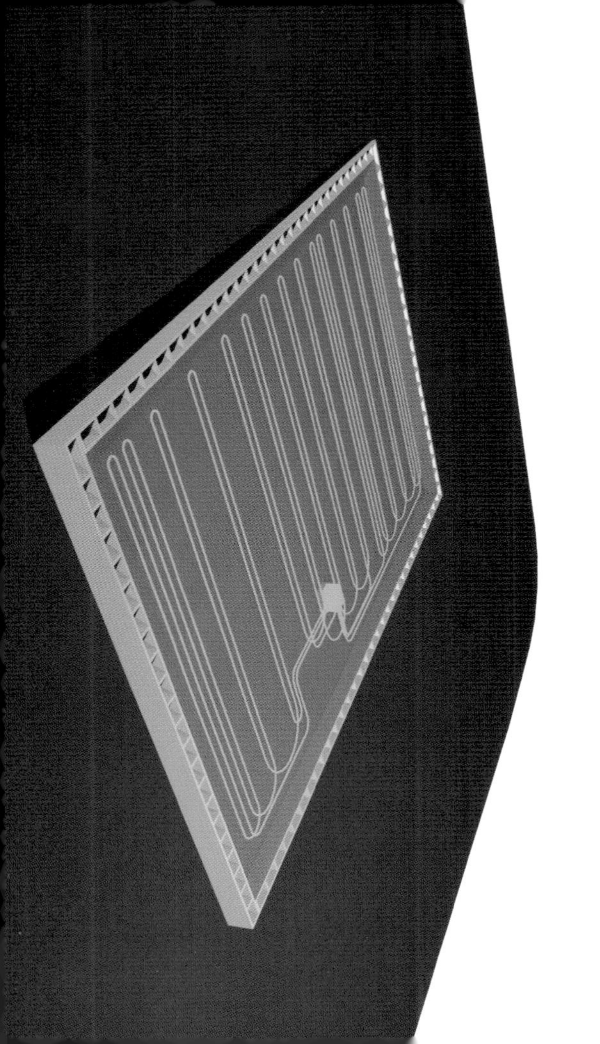

FROST-PROTECTED SHALLOW FOUNDATION

STEPS IN CONSTRUCTING A FOUNDATION

Other things being equal, you should orient your house with its longest and most window-filled wall facing slightly east of south (for sites in the Northern hemisphere). With this orientation, the low winter sun warms the interior of the house through the windows early in the day when the house most needs heat. Once the interior thermal masses (stone or wood floors, for example) are charged with heat, they will continue to warm your house on into the night hours. In the summer, with the sun at a higher angle, its rays, blocked by the roof overhang, will not overheat your house. Windows should be less numerous on the east and west walls, and care should be taken to avoid the glare of the rising and setting sun. Because glare is not a problem from the north, northern light, generally diffuse, is good for lighting buildings during the day.

Red Feather typically builds where winters are severe and thus its houses need the full advantage of straw's excellent insulation performance. The shallow frost-protected foundation we advocate (and describe in this chapter) is geared for the extreme climate of the Great Plains.

Foundation design depends on several factors: local climate, soil composition, building size and weight, and exposure to wind and earthquake forces. You will need to talk with builders in your area, and consult local building codes to determine what kind of foundation is most appropriate for your site. However, do not attempt to design your own foundation. This is work for professionals and the future of your building rests upon the ability of your foundation to do its work—that is, to support the load of the building and provide a strong, dry platform.

Note: Sections of this book that relate to foundations and electrical and plumbing work are quite technical. The point of including these sections in a handbook is not to discourage you. Rather it is to allow your contractor to understand how his/her work fits into your overall straw bale project. The rest of the book is designed, however, for individual owner/builders or community work groups.

Concrete is a readily available and useful building material. However, roughly 7% of greenhouse gas emissions worldwide stem from the manufacture of Portland cement—a

Opposite: Let the site tell you where to build. Build where trees and hills will protect the walls from winds and driven rain.

Right: Access to utilities is an important consideration when choosing a site. On rural sites a bit of trenching is typically required.

major component of most concrete mixtures. Given this negative environmental impact, Red Feather tries to keep the amount of concrete in its structures to a minimum, and advocates that you use concrete mixes containing fly ash, a waste by-product of the coal industry which can replace Portland cement to a great extent. Ask your concrete supplier early on about the possibility of using a fly ash mixture.

The shallow frost-protected foundation recommended below benefits your project in several ways. Less digging means less money, work, and time as well as less concrete, all of which minimize disturbance to your topsoil and site vegetation. However, whatever foundation system you choose, be sure that it provides a level and appropriate base for your bale walls and, most importantly, that it keeps your bales protected and dry.

Frost-protected Shallow Foundation

The way a building meets the earth varies greatly with respect to site, climate, and seismic zone. The notes presented here suggest an approach for the shallow frost-protected foundation that we recommend for cold climates. In frost-free regions, this foundation is simpler to construct because much of the insulation and the entire insulation skirt may be omitted.

Note: Design and construction professionals should be engaged to engineer and build your foundation.

A shallow frost-protected foundation is made of concrete and rigid foam insulation. A stem wall is built in the typical manner (in warm climates a simple foundation will serve). In cold climates, rigid foam is applied to the exterior of a thickened-edge foundation sized according to structural demands, thereby saving the expense, work, and material required to extend a foundation below the frost line. To keep a shallow foundation from freezing, rigid insulation must be employed to keep the frost out. Thus you will need to construct a continuous skirt of rigid insulation around the building starting 1'–0" below the lowest grade and extending outward 6" for each foot of frost depth.

Formwork is removed from the completed frost-protected shallow foundation.

The completed slab awaits straw bales.

The first steps of construction are forming the foundation and placing reinforcing steel.

Insulation skirt in place, the slab is prepared for backfilling.

Steps in Constructing a Foundation

1. Excavation
Start by removing topsoil as required for the foundation. Pile topsoil carefully in a convenient location to spread back around the completed building. Grade a level area down to undisturbed soil for the building's footprint. Dig out trenches for the foundation slab and thickened edges. Place an even 4"–6" layer of washed gravel, or screened and washed crushed rock (3/4"–1") in the trenches. Provide for electrical and water utilities within the interior of the foundation. Plumbing should never be run in the bale wall. It is best if all utilities come up through the floor slab.

2. Place Forms for the Stem Wall
Place forms in the trenches, and make them level and plumb to the dimensions of the foundation plan.

3. Place Steel Reinforcing (Rebar with recycled content above 95% is available):
Follow the foundation plan and details to place steel reinforcing (rebar) components within the foundation forms.

4. Pour Concrete
Pour concrete into perimeter formwork and give it time to cure.

5. Place Slab Underlayment
Check that interior area is level. Place and grade gravel. Add sub-slab insulation, as required for cold climates, under 6 mm polyethylene sheeting.

6. Place Slab Reinforcement
Place slab reinforcement (metal mesh on top) as per foundation plans.

7. Place Radiant Floor System Loop (See Appendix E)

8. Pour the Slab
Be sure to maintain an insulating thermal break between the slab and the perimeter stem wall.

9. Allow for Proper Curing
Foundation must cure before proceeding to the next step (chapter 2).

To form the stem wall itself we have had good luck using permanent, insulated form systems such as the I-Form ICF (Insulating Concrete Form), a hollow-core, poly-styrene, modular form system produced by Reward Wall Systems of Omaha, Nebraska. The I-Form has the distinct advantage of being a recyclable and non-toxic product that is produced without the emission of harmful hydrochlorofluorocarbons (HCFCs) or chlorofluorocarbons (CFCs).

For insulation skirt materials, be sure to use the correct form of insulation. Horizontal insulation for frost protection must be XPS (expanded polystyrene). Vertical insulation may be XPS or EPS (extruded polystyrene) in thicknesses required by the local climate, and specified for the building in question. The greenest (most ecological) of rigid foams currently available is perhaps the molded XPS manufactured by Insulfoam. It is fabricated without HCFC emissions, and is intended for use below grade. More information on frost-protected shallow foundations can be found on the National Association of Homebuilders' website: www.nahb.org.

When integrated with the radiant floor system presented in Appendix E, the shallow frost-protected foundation described above has the ability to heat a straw bale building more than adequately. However, a redundant heating unit such as a wood stove can be added for those who would like an alternative method of heating. With a low-maintenance radiant system in place, a straw bale building is easy and inexpensive to heat even in the most extreme winters. The savings created by a super-insulated building coupled with an efficient radiant heating system reduces the long-term cost of a home remarkably. We have heard that families living in Red Feather straw bale homes save up to 400% on their winter heating bills compared to the cost of heating light-frame wood homes or HUD housing units. Such a savings is capable of dramatically changing a family's financial situation.

Your Roof Bearing Assembly (affectionately known as RBA) is a hollow box beam that you will build with plywood panels on the top and bottom, and wood I-beam joists (also known as TJIs) on the sides. This assembly gives rigidity to the upper bale wall, provides an even-bearing surface on top of the bale wall, and a structure to which you can attach a conventional roof framing (like pre-fabricated wood trusses), and distributes the load of the roof and ceiling components. It also connects the roof to the foundation, and serves as a continuous member spanning door and window openings. A strong, straight, and uniform RBA is a critical component of your straw bale house.

CONSTRUCTING THE RBA

STEPS IN CONSTRUCTING YOUR RBA

There are several ways to construct an RBA. For smaller structures with lighter loads, it may be as simple as a "ladder" built of 2x4 members that is then placed on top of your bale wall. Several years ago, we adopted a box beam style of construction for larger buildings, and more recently we have been using I-beam joists (commonly used for floor construction) for the structural sides and internal blocking of our RBAs. An RBA built of I-beam joists is recommended because of its light weight and increased strength compared to an RBA made with solid lumber. In addition, I-beam joists can be purchased in almost any length. Moreover, they are a sustainable building product since they are fabricated without the need to sacrifice old-growth, large-diameter trees.

Constructing the RBA

Despite its position above the straw bale wall, the RBA is the first thing you will build once your foundation is complete. It is crucial that the dimensions of your RBA match those of your foundation slab exactly if your walls are to be straight and plumb. So it is best to use your foundation slab as a template for laying out and constructing your RBA. Later in the construction process, the slab floor will be cluttered with other elements of construction, so build the sections of your RBA on the slab and then set them aside for later installation.

Note: Make sure to have enough people on hand to help move the RBA sections safely. It is best to assign one person the responsibility of directing the group when moving RBA assemblies.

Opposite: Use the foundation as a template for RBA construction.

Right: RBA blocking being nailed in place

Steps in Constructing Your RBA

1. Layout Construction Lines on Slab
Using a chalk-line, measure the width of the RBA from the slab edge and snap guidelines on the slab itself.

2. Cut Joists
Cut I-beams to size and lay the pieces on the established chalk-lines.

3. Cut Spacers
Cut lengths of scrap as spacers (blocking) that will run perpendicular to the sidewalls of the RBA.

4. Nail the Box Beam Together
Fasten the I-beam portions of the RBA together providing perpendicular I-beam blocking at regular intervals; the interval should match that of the trusses that will bear on the RBA. Verify the spacing with the roof design drawings and make sure that each truss will rest above blocking. Build straight RBA sections that are not too large to handle later during RBA raising. At this point the RBA will look like a number of ladders each awaiting the attachment of its plywood sheeting.

5. Rip Plywood Sheeting and L Sheets
Cut strips of plywood wide enough to serve as the top and bottom sheets covering the I-beam ladders. Also cut L-shaped corner connector pieces and set them aside. These pieces will be used as connectors when you install the RBA on the top of your bale wall. When 24" wide 3-string bales are used there is little waste of plywood for the sheets are cut to 24" widths and the L-sheet remainders may be used to cover the straight portion of the RBA.

6. Sheet One Side of the Box Beams and Fill
Fasten a plywood sheet to one side of the RBA ladders. Then turn these ladders over and fill them with loose straw flakes for insulation.

7. Reposition Beams on Slab
Set the straw-filled box beams back on the construction lines you drew on the slab, and line them up carefully.

8. Sheet the Topside of Each Beam
Nail plywood tops onto the insulation-filled ladders.

Placement of RBA L-sheet assembly. Note that L sheets are not fastened until the RBA is atop the wall.

Use the foundation as a template for RBA construction.

9. Number Sections and Slab, Stack Completed Sections Aside

Write "key" numbers on the various RBA sections and write the same number on the slab area on which each piece was formed. This is necessary since it will enable you to put the right RBA sections on the part of the foundation slab where they were laid out and constructed. Finally store the RBA pieces, and clear the foundation area to begin the construction of your bale wall.

Power Tools

When Red Feather volunteers arrive on site for a build, the first task is the preparation of base plates. Base plates provide a stable and dry place on which you will set the first course of straw bales. These plates consist of strips of lumber laid out around the foundation edges. The outer base plate is installed flush with the outside face of the foundation. The inner plate, which may contain a chase for electrical wiring, supports the bale wall. You will fill the space between these inner and outer plates with rigid insulation that will serve as a thermal break between the interior of your home and the outdoors. Elevating the base of the straw bale wall from your foundation slab in this way protects the bottom of the bale wall from moisture that might collect on the slab in the event of, for example, a plumbing failure.

BUILDING THE BASE PLATES

STEPS FOR BUILDING BASE PLATES

With the bale wall raised by these base plates, there is little chance that the bottom course of bales will ever be sitting in standing water, as they might be if the first row of bales were placed directly on the foundation. Any moisture that finds its way into the bale wall from the sides or top will drain out through the gaps in the rigid insulation you will place between the plates and then be directed outdoors through weep holes—shallow saw cuts—that are cut into the bottom of the outside base plate. Base plates, in short, keep your bales high and dry above the finish floor. They also serve as important points of attachment for later construction. Lath, base screed for stucco coats, and the metal coil straps that connect the plates structurally to the RBA all will be fastened to the exterior base plate. The interior base plate functions not only as a channel for electrical wires, but also provides an attachment point for base screed, and eventually base trim.

Building the Base Plates

Lay out your base plates using the same chalk lines you used to construct your Roof Bearing Assembly (chapter 2). If the lines are not clear, you will need to re-snap them.

Next, locate the door openings. As the base plates do not run through these openings, you will need to identify door positions prior to laying out the base plates. Be sure not to use glues and/or fasteners in doorways if your slab is to be the building's finished floor.

As with RBAs, base plates can be as simple as treated 2x4s laid flat under the bale wall with insulation placed between them. However, in larger structures, it is best to use a double layer of pressure-treated 2x6s for your exterior plate and, for the interior one, a built-up assembly of pressure-treated lumber in the form of a C-shaped channel opening to the inside. This channel will be used to contain the electrical conduit.

Once the base plates are laid out and positioned, a waterproof membrane is attached. Then you can fasten the plates to the foundation with a ram-set, or bolt and pin them to the slab with a roto-hammer and wedge anchors or rowel pins.

After you have attached them to the slab, drape the interior base plates with a layer of waterproof membrane, and fasten it in place to protect the wood plates from contact with later stucco work.

Next attach the exterior base plates to the slab (over waterproof membrane) and fill the space between them with strips of rigid insulation left over from building the foundation. Cut these insulation strips to the height of the base plates and install them vertically in order to maximize the number of seams that will be able to convey water down and away from your straw bale wall.

Traction nails help lock the first course of bales in place.

Complete work on the base plates by providing traction nails to prevent the bales from moving from their seats atop the base plates. We recommend nailing 20 penny galvanized nails about half-way into the base plate (space at 8" on center and stagger the nails between the inner and outer plates). The exposed portion of this nail array will keep the bottom of your bale wall firmly in place.

Note: Pressure-treated wood must be used for all wood directly in contact with concrete foundations and where wood may be exposed to water. The chemicals used to make wood rot-resistant have typically made it hazardous for those who used it or breathed its toxic sawdust. Thankfully, commonly available, pressure-treated wood products now exist that are non-toxic and therefore far better for the environment and carpenters alike. Make a point of selecting non-toxic pressure-treated wood products in order to enjoy a healthy home and a safe construction site.

The layered base-plate assembly: 2x4 plates over waterproof membrane over frost-protected foundation

Powder-actuated fasteners hold the base plate to the slab.

Weep holes allow the base plate to drain.

Steps for Building Base Plates

1. Establish Construction Lines
Clear the area around the edges of your slab, and re-snap the chalk-lines you used in constructing the RBA if necessary.

2. Layout Door Openings
Mark the location of all doorways around the perimeter of the slab. These openings will not receive base plates.

3. Layout and Cut Pressure-treated Plate Stock
Cut base plate sections to length and place the pieces in their correct locations.

4. Assemble the Interior Base Plate
Nail and glue the component pieces of your C channel base plates together.

5. Attach the Interior Base Plates to the Foundation Slab
Cut and/or position base-plate sections so that they fit together at connections and corners. Then attach the plates to the concrete slab using one of the fastening methods listed above.

6. Moisture Protection
Drape a continuous piece of waterproof membrane over the top of the interior base plate and then outward to the edge of the foundation. This layer will serve to keep any moisture that develops in the bale wall from entering the building and will guide it out through the weep holes produced in the next step.

7. Moisture Escape
Make weep groove cuts across the bottom of the exterior base plates to allow any moisture within the wall a means of escape. Using a circular saw (set at a cutting depth of ¼") cut ¼" deep, by ¼" wide weep grooves across the bottom of the exterior base plates every 24" along its entire length. This is a crucial step and will prevent the bale wall from trapping moisture.

8. Attach the Exterior Base Plates to the Foundation Slab
As with the interior base plate installation, cut and position exterior base-plate sections so that they fit together at connections and corners. Then attach the plates to the concrete slab. At this point, the waterproof membrane should be located over the top of the interior plate and under the exterior plate. This will prevent moisture from entering the house and ensure that it will drain outward and not be trapped within the wall.

9. Cut and Fit Rigid Insulation

Cut strips of rigid insulation to fill the gap between the base plates. Use scraps of insulation from the foundation or other work if possible. It is best to place these small pieces of rigid insulation perpendicular to the base plates so that there will be numerous places to allow water to escape.

10. Cover Any Exposed Plate Tops with Waterproof Membrane

Staple roofing felt (tarpaper) over the plates to prevent them from coming in contact with stucco. On the interior side of the plates, the felt should run just to the top edge of the wood channel assembly built to hold the electrical conduits. Make sure the felt covers only the wood plates and not the insulation between them.

11. Traction Nails

Nail traction nails halfway into the base-plates as per above description.

Detail of base plates with insulation

CHAPTER 4 | WINDOW AND DOOR BUCKS

Before you raise the bale wall, you will need to address several carpentry tasks. At each corner, attach a guide that is as high as the bale wall will be. These corner guides will assure that the wall is plumb and square to its foundation. Additionally, rigid wood door and window frames—known as "bucks"—should be constructed prior to the bale wall raising. Door bucks are erected along with the corner guides while window bucks are set aside

BUILDING WINDOW AND DOOR BUCKS

STEPS IN BUILDING WINDOW AND DOOR BUCKS

Corner guides are easily constructed by joining two straight boards (typically 2x6) of the same length together to make an L-shaped piece. Attach the bottom of one of these four L-shaped pieces to the base plate at each corner, and then diagonally brace each on both sides making sure the corner guides are plumb. Since these guides will be taken down after you complete the walls, be sure to use fastening techniques that facilitate easy removal.

Window and door bucks are wood frames constructed of plywood and 2x4 lumber that provide sturdy openings in the bale wall as well as solid frames for the attachment of windows and doors. Structurally, bucks serve to protect doors and windows from expansion or movement of bale walls. Bucks float within the wall and transfer loads around the window and door frames to the wall and foundation.

When possible, windows and window bucks should be sized to the length of a bale to facilitate the straw bale wall building process. However, the size of salvaged, discounted or donated windows may not correspond to bale dimensions. In any case, each buck needs to be sized properly for the window or door assembly it will hold. Window bucks are built and set aside until the bale wall reaches the window's sill height—typically atop the second course of bales.

Install door bucks before any bales are stacked in the walls and attach them directly to the base plates. Both window and door bucks provide nailing surfaces for attaching lath around wall openings, and create strong frames for installing window and door units. These bucks also provide a place to attach stucco screeds, and any trim or millwork that surrounds such openings.

Sturdy and plumb corner guides lead to straighter walls and easier construction.

Built beforehand, bucks are quickly located during the wall raising.

After you have completed the foundation and laid out the base, you will install door bucks and corner guides.

Building Window and Door Bucks

The first step in building bucks is to establish the actual size of your windows and doors. If manufactured windows are used, the rough-in dimensions will be available from the manufacturer's catalog or cut sheet. If the windows are on site they may be measured directly.

Once the size of rough openings for doors and windows are known, you can calculate a buck's interior dimensions. For windows, add 2″ to the rough dimension for both height and width: for example, a 30″ square window requires a 32″ square buck frame. Door bucks are sized 2″ larger than the rough opening dimension side to side, and 1″ in height, so if the rough opening for a door assembly is 3′–0″ wide by 7′–0″ high, the buck interior dimensions should be 3′–2″ wide by 7′–1″ high. It is critical to get the correct window and door dimensions prior to building the buck frames, for once the bucks are locked into the bale wall with stucco it is extremely difficult to make changes.

Construct bucks with a double 2x4 frame sheeted with a 12″–16″ skirt of 3/4″ plywood.

Completed window and door bucks are in place.

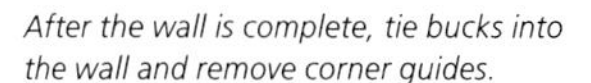

After the wall is complete, tie bucks into the wall and remove corner guides.

Steps for Building Window and Door Bucks

1. Establish the Exact Sizing of the Windows and Doors to be Installed

Verify rough opening dimensions for each window and door, and calculate the slightly larger buck dimensions according to the procedure outlined above.

2. Build the 2x4 Frames

Cut 2x4 frame pieces to the size of the buck dimensions and assemble the frames. Overlap frame pieces when possible for added strength. Be sure that the frames are square when complete.

3. Rip ¾" Plywood Skirts

Cut plywood strips to skirt the 2x4 buck assemblies. Then cut these strips to length using the pre-built 2x4 frames as a guide.

4. Screw or Nail Skirts to 2x4 Frames

Screw the plywood pieces to the outside of the 2x4 frame holding the plywood edge flush with the frame on one side (this side will be installed facing the exterior of the wall).

5. Screw or Nail Skirt Corners Together

Fasten the free corners (those away from the 2x4 frame) together to add rigidity to the buck frames.

6. Square and Install Temporary Bracing

Make sure the completed frames are square and then screw temporary bracing (scrap wood 1x4 or plywood) to keep the frame square until it is tied in to the bale wall.

7. Store Bucks

Set the completed bucks aside for later installation within the bale wall.

Construct and brace window bucks before you need them to expedite the raising of the bale wall.

The straw bale wall is the heart of a Red Feather project and often the construction sequence that is least familiar to experienced builders. Bale building is very intuitive and lends itself easily to construction by groups of volunteers and community members. By adopting this fun and accessible technology, a motivated group of people can stack the walls of a home in less than two days. First bales of good quality must be sourced; then the community-friendly bale wall raising event may begin.

PART 1: CHOOSING AND STORING BALES

STRAW BALES SHOULD BE

BALE STORAGE

BALE COUNT

PART 2: STRAW BALE WALL BUILDING

CUSTOM BALES

PREPARATIONS

LAYING UP STRAW BALE WALLS

CORNER STAPLES

WINDOW BUCK INSTALLATION

BALE PINNING

PATCHING GAPS IN BALE WALLS

PART 1: CHOOSING AND STORING BALES

For the load-bearing structure described in this handbook, the greater stability (from increased width) and higher compaction of three-string bales make them the best choice. These bales typically measure about 22" wide by 15" high by 48" long. Be advised that some three-string baling machines produce bales that vary slightly in height, and that the length of the bales in any batch will probably vary by at least four inches. Red Feather generally favors the wheat straw common to the Northern Plains or rice straw when it is available. For projects in the Southwest, the wheat straw bales made by Navajo Agricultural Products Industries (NAPI) are an excellent choice. NAPI harvesters are very knowledgeable about the requirements of building-quality bales and typically have a large quantity of bales in dry storage. Other straws can also be used successfully, but as these have different properties from wheat or rice straw, you will need to do some research before considering them for use in building. First bales of good quality must be sourced then the community-friendly bale wall raising event may ensue.

Like lumber, straw varies greatly in its quality and composition, but in straw bale construction there is nothing comparable to the timber industry's method of visual grading. Given the current absence of such standards, consider the following criteria in choosing your bales.

Straw Bales Should Be:

Dry. Straw used in bale buildings should be harvested during dry weather and stored in a dry place where it can be thoroughly protected from rain, and from water on, or in, the soil. It is also important to have good communication with the farmer growing the straw in order to know the history of a given lot. In any case, you should measure your bales with a moisture meter. Ideally bales should have a moisture level at or below 10%.

Dense. Bales need to be compressed to a higher degree than many farmers are accustomed to producing. Explain that the bales you are seeking will be supporting the weight of the roof of a house and therefore need to be really dense and sturdy. Ask the farmer you contact to set his baler to a medium-high tension. If you are lucky enough to find a farmer who is both patient and somewhat technical, you can request bales that have a dry density ranging between 7 and 8 pounds per cubic foot.

Well Tied. To hold a dense bale together, Red Feather relies on strong bale ties. Traditional baling wire (metal) is acceptable, but it may condense moisture inside the bale wall and possibly rust. Baling twine made from natural fiber will degrade over time and may rot, and is therefore not recommended. The best (and most commonly available) building bales are tied with polypropylene twine that is treated to resist degradation from exposure to sunlight.

Pure. The best bales will be free of seeds, weeds, and any foreign objects. Ideally bales should be composed of straws that are long, thick, and uncrushed. Bales made up of short or shattered straws contribute dust to the building site and probably have less structural integrity.

Fresh. It is fine to store bales for a few months in an appropriate manner, but it is better to use recently harvested bales. This limits opportunities for water infiltration, and allows you to be confident of your bales' history. A deep yellow color indicates that a bale is healthy and recently harvested. Bad smells or discoloration indicate that a bale has been exposed to water and is beginning to compost.

Consistent. A batch of bales for building should be similar in size and shape. Bales that are of the same height and width will be easier to lay up in courses, and will result in a better building with less effort. Bale length can vary considerably, however, but there are distinct advantages to using bales that are about twice as long as their width. Such bales will work well when they are stacked in running bond.

Bale Storage

Unless your bales can be delivered the day you start constructing your walls, you must properly store them on site. Even with the best of scheduling, you must be prepared to protect the straw should rain come. An empty building nearby is the easiest solution. If no such structure is handy, Red Feather typically stacks straw bales on wood pallets and keeps them protected with sturdy tarps.

Bale Count

A bale building requires an exact number of bales. However, some of the bales you receive will be sub-standard or will have fallen apart in transport. You will also need extra bales for staging, for insulating gaps around openings, and for the Roof Bearing Assembly (RBA). To provide for these extras, and for unforeseen contingencies, we recommend that you order 15% more bales than you need for the walls alone.

A good bale does not deform when you pick it up by one string.

Have plenty of tarps on hand in case of rain.

adidas

PART 2: STRAW BALE WALL BUILDING

After the RBA segments have been built and the base plates, the corner guides, and the door bucks have all been installed on the foundation, actual straw bale building can begin.

Anyone who has played with LEGOs has a leg up on understanding the straw bale construction process. Basically a full bale functions as an 8-bump (rectangular) LEGO piece and a half bale as a 4-bump (square) piece. Any LEGO-savvy kid intuitively knows the difference between "stack bond" (bales stacked directly on top of each other) and "running bond" (staggered or overlapped bales). Whenever possible, stack the bales in your walls in running bond.

Custom Bales

There will be a need for half bales and other custom bales of various lengths in even the most simple straw bale building layout. It may prove useful to make a number of half bales in advance, as these bales will be needed at door and window buck locations. Most other custom bales will have to be made after a length measurement is provided by one of the bale-laying teams. Creating several custom-bale teams may be helpful so that the relatively slow process of custom bale making does not delay work on the walls.

To make a custom bale, begin with a full length bale, but use the less perfect bales in your stock for this purpose. Bales with missing strings, damaged ends, and other imperfections are good candidates. Keep the best end of a custom bale intact; customize the worst end. Then, using a bale needle—a long piece of metal flat stock notched at the tip to hold a knotted end of a piece of bale string—thread a piece of string through the center of the custom bale long enough to tie it. You will be making three such loops. The idea is to get these new loops in place before you cut the original bale loops. If you do this right, you will have a well-formed custom bale and some leftover straw flake. Keep both dry and in good condition until needed.

Opposite: Get friends to help you raise your walls.

Right: Weed whackers, hand saws, rotary grinders, hedge clippers, and chainsaws find new use in shaping custom bales.

Far right: Check work often to ensure walls that are both plumb and square.

If you have to use a good bale to make a partial bale, and a reasonable portion of the bale is left over, it may be worthwhile to make this remainder into a mini-bale. Measure the length of this orphan bale, write the length on a piece of paper and slip it under one of the mini-bale's strings. Such orphan bales, lined up according to length, will make it easy for a bale-laying team to see if a mini-bale already exists to fill a particular gap.

Preparations

It is helpful to draw diagrams showing the position of the bales that will make up the building well before construction begins (see Appendix B for examples of bale diagrams). Bale diagrams are useful in part because you need to have a good idea of the number of bales required for your project—including extra bales for custom bales, gap insulation, and bale scaffolding. A bale diagram is also useful as a reminder of door and window locations. With a diagram, a sufficient quantity of bales, and pre-built window bucks on hand, bale walls can go up quite quickly, inspiring morale and reducing the chances of the bales getting wet. However, before you start stacking bales, use your level to make absolutely sure that your corner guides are plumb. If the corners of your bale wall go up wrong, you'll have little choice but to take some bales down and start over. One of the appealing aspects of bale building is that mistakes can be remedied quite easily until the RBA and the roof are added to the building (all that weight makes adjusting bales difficult). In other words, regularly check your drawings, and that your walls are rising plumb, so that you will not waste time fixing avoidable mistakes.

Laying up Straw Bale Walls

The placement of the first layer of bales (those that rest directly upon the base plates and are held in place by the traction nails that you've driven in partway) is most important. The layout of the first course sets the pattern for the bales to follow. Care must be taken to insure that this first course lines up with the base plates and the foundation. If these first steps are done properly, the rest of the wall raising will go quickly and a better bale wall will result.

The first bales are placed at the corner guides and at the outside of the door buck assemblies. Line the bales up at the exterior edge of the base plates and continue working along the first course until it is complete. On load-bearing straw bale buildings, you should not push the bales tightly to the plywood face of the door bucks. Instead, leave a gap of about 1" for the bale wall to fill when it is compressed by the weight of the roof. Compression will expand the bales into this gap and any remaining gaps can be stuffed after the wall has fully compressed. The last bale in any wall section should be measured against the gap which it is meant to fill. It may fit perfectly, or it may need to be replaced by a smaller custom bale. If the fit is close but not quite right you can take a measurement and search among your supply (or your row of mini bales) for a bale that fills the gap, for a batch of bales usually varies considerably in length. Any remaining small gaps should be filled with straw flake; if

Stack bales in alternating courses to avoid long, vertical seams.

The first course rests on the insulated toe-up.

The first course is placed with small gaps between each bale.

a gap is larger than about 4", move an adjacent bale half the gap distance and stuff the two resulting smaller gaps with loose straw.

Once your first course of bales is in place, give the wall a close inspection. Add straw to any gaps or depressions you find, and then lay the second course in running bond—that is, make sure each bale in the second course rests on two bales of the first course, bridging the seam between these bales.

Window bucks typically rest on top of the second course of a bale wall. So, referring to your drawings, measure off and then locate your pre-assembled window bucks on the second course after it is completed and checked. Red Feather also starts adding corner staples (see below) at this point, placing one staple per corner. You will need four corner staples for each course except for at the first course which is held firmly in place by the traction nails in the base plates.

After you complete each course, jump up and down on the bales to "seat" them, and to compact the bale wall. It is helpful to attach a 4' level to a straight 8' long 2x4 in order to square and level the successive course of bales with the corner guides. Using a level to align bales as they are stacked upward will insure a straight and beautiful wall.

Corner Staples

Bend lengths of #3 rebar to the shape of a large staple. The size of your corner staples will depend on the size of the bales involved. For a typical 3-string bale wall laid flat to achieve a wall thickness of about 24", corner staples should be about 24" wide with right-angle legs that can stick 10–12" into the wall, so you will need to start with something close to a 4' length. Use these corner staples to connect the two bales that make up each corner of a bale course, and reduce the likelihood they will be bumped out of position. Together with bale pins (see below), corner staples add stability to the wall while it is being built, making it easier to walk on, and harder for a strong wind (or an RBA team) to knock it down.

Window Buck Installation

As mentioned above, window bucks are usually installed on top of the second course of bales. This depends on the window, of course, as small windows are sometimes placed high in the wall on the third or even the fourth course. A floor-to-ceiling window is also possible, but needs careful structural consideration.

Once a window buck is in place, use shims as needed to level the bottom part of the buck. You should install a temporary brace to hold the buck safely in position until it is secured in the wall by pins or some alternative. The first bales of the third course are now placed at the corners and on both sides of the bucks leaving about a 1" space (as with the door bucks) to

Bent rebar staples connect bales at the corners.

Leveling bucks

Half bales are often needed at window and door bucks.

allow for expansion.

When the window bucks are completely surrounded by bales, hammer short pieces of rebar (8″–10″ long) through pre-drilled holes in the window buck thereby securing the buck to the wall. If the bales are very compact, it may be necessary to grind one end of these pins to a point. Do not drive the pin all the way through the hole in the buck. Leave a few inches of the pin visible. Once the RBA is in place and any final adjustments have been made to the bale walls, these pins can by pulled out (vise-grip pliers work well for this task) to allow for the final positioning of the buck within the opening. At this point, drive the pins into the bales again. For this final setting of the pins, drive them in so that only about ¼″ of the pin is visible from the inside of the buck assembly. This will provide secure attachment. The end of the pin will be easily covered by the lath and stucco to come.

As walls grow taller, Red Feather uses extra bales to build stepped scaffolding. By laying tiers of bales on top of one another, you can create good temporary platforms to stand on while you work. These straw bale work-platforms are useful for positioning the last few courses of bales, pinning the bales and window bucks, and installing the RBA atop the wall.

Left and right: Extra bales come in handy for temporary stairs and scaffolding.

Hammer short lengths of rebar through pre-drilled holes to connect window and door bucks to bale walls.

Use a sledgehammer and a pipe end to drive rebar pins into a wall.

A bale hammer, built on site from a log section and a strong wood handle, persuasively adjusts bales after they have been stacked.

Bale Pinning

After the fourth course of bales is in place (and then again after the sixth and final course), Red Feather drives straight pins, approximately five feet in length, down through the bale wall for reinforcement. Using a sledgehammer, pound two pins into each top bale. As you will do this after completing both the fourth and sixth courses, the pins will overlap at mid-wall height. Such pinning makes the wall more stable during construction, and may offer structural benefits after the bale wall is fully loaded.

Various types of pins—bamboo, wooden surveyor's stakes, saplings—can be used; however, Red Feather typically uses high-recycled #4 rebar ends salvaged from reinforcing steel left over after building the foundation.

After pinning the final course, you are ready to proceed to RBA installation, but first make a thorough inspection of the wall, patching gaps and seams with loose straw. Also at this time use a bale hammer to encourage protruding bales back into the plane of the bale wall.

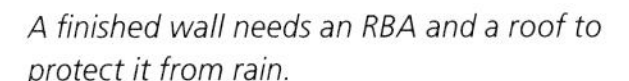

A finished wall needs an RBA and a roof to protect it from rain.

Patching Gaps in Bale Walls

Even when much care is taken to make bales snug and straight, spaces between bales will occur, and will require attention. Add loose straw to gaps to assure a consistent layer of insulating straw throughout the bale wall. Loose straw on the deck is a safety hazard, being both slippery and flammable. By periodically gathering the loose straw that falls out of the walls during the building process, you will have a constant supply of straw to stuff gaps between bales. Red Feather keeps a few large plastic garbage bags handy for this purpose. When the wall is complete, you will fit these same bags—again filled with loose straw—in the openings between the top of a window buck and the bottom face of an RBA to provide consistent insulation as well as backing for the lath that runs from the top of the bucks up to the RBA. You must finish all straw stuffing before any lath is applied to the walls.

It is important, however, to avoid overstuffing gaps, openings, and seams. Too much stuffing is as bad as not enough, and perhaps worse, because overstuffing can cause bulges that make applying lath difficult. Work toward an even, uniform surface with no big bulges or indentations (except for the occasional decorative niche). However, do not be too rigorous about creating a very flat wall, since part of the beauty of a straw bale wall comes from its picturesque surface irregularities.

Raising a bale wall is usually a satisfying and (if you are successful in gathering together some helpers) a fast-moving experience. There is a danger, however, of losing all the work that went into a bale wall; if rain gets to your bales, you will be doing another bale-raising after removing the soggy bales and purchasing replacements. So protect your wall from water at all cost. Have tarps handy until you can build a good roof.

For ease of construction, windows are sized to bale dimensions.

Tarps protect the bale wall from the elements.

CHAPTER 6 | **ABOVE THE STRAW BALE WALL**

Installing a Roof Bearing Assembly is perhaps the most daunting and exciting part of a Red Feather project. Because of size and weight, you will need many able bodies and considerable cooperation to place pieces of the RBA on top of your bale wall. Given adequate preparations in terms of group organization and safety, this work will go quickly and safely and represents a very satisfying team-building exercise for all involved. Once the RBA is in place, the installation of trusses, gable-end framing, and roofing may begin.

INSTALLING AN RBA

STEPS TO INSTALLING AN RBA

ROOF FRAMING AND TRUSS INSTALLATION

BARGE RAFTERS AND SOFFITS

GABLE ENDS

CEILING DRYWALL INSTALLATION AND INSULATION

ROOFING

OLD NAVY
Eagles
Soccer

Installing an RBA

Since you have already constructed the sections of your RBA (see Chapter 2), you will not need much in the way of materials for this step. However, you will need a fair number of extra bales, since Red Feather typically makes a temporary, stair-type, straw bale scaffold beside one or more of the exterior walls. These bales provide a tiered work platform that can be easily moved from one wall to another, and will allow you and your friends to carry long RBA sections and place them on top of the wall. The more people you can round up to help you do this, the better.

Starting with one of the larger sections of your straw bale wall, position a few "wall walkers" atop the wall (preferably folks who are comfortable walking on top of the walls), and ask several other people to act as spotters. The number of spotters and walkers depends upon the size of the section being lifted and the strength and confidence of the spotters. There should be at least two people in each role, but several more may be required for larger RBA sections to insure coordination and safety. Each spotter needs to have a 2x4 about 10′ long to steer and push sections this way or that while they are being positioned. Those not fortunate enough to have drawn a better job get to carry the sections up the bale scaffold steps and lift them into place.

When a large number of people are working on installing an RBA, it is best to designate one spotter as the only person directing traffic. Early experiences with RBA raisings were reminiscent of a noisy group of Vikings trying to ram a castle gate open with a massive log. It is easy for everyone to get excited when moving objects of this size, but if folks get moving too quickly, safety will be compromised. In fact, it is possible to knock a bale wall down (and the people standing on top of it) with a runaway section. But if everyone works together under the calm direction of one lead spotter, each section can be walked up a ramp and slid carefully in place.

Once the sections are up and positioned, nail the corners together, and then install a pre-cut plywood L sheet at each corner.

After you have checked that the RBA as a whole is square and level, use it as a straight edge to align your bale walls. By walking along the top of the RBA and sighting down to the foundation line, you will be able to tell which bales are sticking out of line and need to be adjusted with a few strokes of a bale hammer.

To complete the RBA installation, use a long spade bit to drill through the bottom layer of the RBA plywood. After drilling, verify that the RBA and the bale wall are in alignment, and then pin the RBA to the top courses of the bale wall with a 3′–0″length of #4 rebar bent into a 4″ L at the end to keep it from going completely through the hole.

Steps to Installing an RBA

1. Call Everyone Who Owes You a Favor

A big group of workers is invaluable for lifting bulky pieces of an RBA.

2. Build Tiered Bale Scaffolding

Use straw bales to build steps up the outside of the bale wall section where you are installing an RBA section. When you are finished on one side, then reassemble the bale stairs on the other sides as needed.

3. Assign Work Duties

Most of your work party will be carrying and lifting RBA sections, but assign several people as wall walkers and spotters. Also appoint a lead spotter who will be in charge of guiding the RBA team as you move large sections into position.

4. Put the RBA Sections in Place

Lift each pre-assembled portion of your RBA into position under the direction of the lead spotter.

5. Check Plumb and Level

Use a bale hammer or the side of a sledge hammer to position the RBA squarely above the foundation.

6. Fasten Corners

Nail the corners of the RBA sections together.

7. Fill Voids

Fill any voids in the corners of the RBA with loose straw insulation.

8. Install Plywood L Sheets

With a square, a level, and persuasive tools, again verify that the RBA is as square and level as possible, and then, using pre-cut L-shaped sheets, fasten the RBA pieces together to form a rigid assembly that runs around the top of the bale walls.

9. Align the Bale Walls

Use a bale hammer to adjust bales that stick out from the building's interior and exterior. It is helpful to sight along the vertical plane made by the exterior edges of the foundation and the RBA in order to spot protruding bales.

10. Pin the RBA to the Walls

When the RBA is fully square and in position, fasten it to the walls by hammering #4 rebar pins down through holes you have drilled in the RBA's plywood panels.

Short sections of an RBA may be raised at once if enough people are on hand. Notice the wall walkers ready to receive and place RBA pieces atop the wall.

Build straw bale scaffolding along walls where RBA sections will be placed.

Slide trusses onto the RBA platform on pieces of lumber set on edge and angled against the wall.

Erect and adequately brace your trusses until you place sheeting over the entire roof.

More than half the trusses are installed, fastened, squared, and braced to the RBA and to each other.

The roof is coming together. Once truss bracing is complete, fascia, sheathing, and gable end work may proceed.

A layer of building felt applied over the gable end allows wood siding or lath-and-stucco treatment to match the walls below.

A gable end frames a view of the Great Plains.

Roof Framing and Truss Installation

For a single-story, load-bearing bale home, trusses are a good idea for many reasons. First, they can be made with smaller dimensional lumber, allowing for a strong, lightweight roof system. Second, the lumber used in wood trusses typically comes from smaller, second-growth, or plantation timber sources. Finally, trussed roofs are quickly erected and cost-effective. For straw bale buildings, where minimizing bale exposure to the elements is all-important, a trussed roof is a good choice.

When working with trusses, use a team approach in moving large sections of material similar to the one you used in raising your RBA. For a simple building, the required roof trusses are quite manageable in weight and size.

Using bales for staging, and a lead spotter to direct traffic, a small group of people can quickly pass individual trusses to a couple of wall walkers on top of the RBA who, in turn, can move along raising and bracing each truss. Use a "hurricane" clip at each side of each truss to connect it firmly to the RBA. These clips act against uplift and winds to hold the roof firmly to the building.

Barge Rafters and Soffits

After the trusses have been raised and braced, you can begin framing the eave and gable overhangs. Install 2x4 lookouts (extensions to support barge rafters) to create eave overhangs at both gable ends of the building of at least 16". On the eave overhangs, snap a chalk-line and then cut the rafters to length. Install fascia trim around the butt ends of the rafter tails. Next, it's time to sheath the roof with plywood. In the overhanging eave section, use material rated for exterior use if you chose an open soffit (an open under-eave area) design.

Gable Ends

Roof underlayment and finish roofing should be applied quickly to protect the building from the elements. While this is being done, frame the gable end walls. Typically Red Feather uses conventional wood framing within the gables. Gable faces may be finished in a variety of ways. Board and batten or another type of exterior wood siding is a good choice. Corrugated metal is another durable option. Stucco may also be applied on the gable ends. If you elect to apply stucco, an expansion joint channel should be installed at the top of the RBA to allow for the fact that straw bale walls and wood-framed gable ends expand and contract at different rates.

Ceiling Drywall Installation and Insulation

At this point it is necessary to install ceiling drywall to help compress the bale walls. You can make ceiling drywall installation easier if you learn several techniques for working with heavy, cumbersome sheets. As with other parts of straw bale construction, it is good to have a number of people around to help.

If your room size allows it, 12′ sheets result in less work and less material to purchase. Fewer drywall joints translate into less drywall tape and mud. Drywall lifts and pulley jacks on rollers, if available, make quick work of hanging ceilings if labor is sparse. Otherwise, construct a few "T-jacks"—that is, scrap lengths of 2x4 lumber shaped into a T such that the top of the T measures 5/8″ (the thickness of the drywall material) shorter than the floor to ceiling dimension. For example, a structure with an 8′–0″ finish ceiling height will require a T-jack that measures 7′–11 3/8″. One person with a T-jack can support a drywall sheet while others attach it to the ceiling framing with drywall screws. Red Feather uses 1 5/8″ standard drywall screws for 5/8″ drywall and 1 1/4″ standard drywall screws for 1/2″ drywall.

Your straw bale wall has very high insulation value, but you must insulate the roof if you want to ensure that the entire building envelope will be energy efficient and that your building will be comfortable. Once the ceiling is in place, put insulation between the rafters of the roof (if the attic will be heated) or between the members of the ceiling plane (if the attic will not be heated). Refer to local codes and climate information to determine what R-value insulation will be required. Red Feather typically uses fiberglass insulation "batts" placed horizontally in the ceiling plane, but many types of insulation are available. For example, good sustainable cotton insulation is manufactured from factory scraps of blue-jean material.

Fire resistant drywall provides a good finish for walls and ceiling. Note the use of a T-jack to support a drywall sheet.

Roofing

After you skin the trusses with plywood, and attach a layer of waterproof membrane to the roof, you are ready to install your roofing. It is important to put a good roof over straw bale walls, and to do so in a timely manner. As stressed before, you must keep water out of the bale walls. A quality roof, quickly built, ensures that your bales will be dry during construction and for the duration of their life.

Red Feather uses several types of roofing on its projects. Sometimes materials are donated, and different regions and climates suggest or demand different materials. One thing, however, is certain: the foundation and roof must keep the straw bales dry or the building will lose its structural integrity, and its right to be called a home. With this in mind, it is important to obtain high quality roofing materials and to control roof construction carefully enough to protect the labor and resource investments your straw bale structure represents.

The design of the roof in this handbook—a simple gable structure with no valleys—is about as straightforward a roof as you can find. The roof's simplicity is important for ease of construction and low cost. In straw bale construction it is important that roof design stay simple because complexities, valleys, and penetrations create opportunities for water leaks as well as rising costs. A roof must also have properly sized overhangs to protect bale walls from driving rain. Overhangs should never be less than 16"—preferably 24"—and should be increased for buildings in rainy regions or for exposures where driven rain is expected.

Roof pitch is another issue you need to consider carefully. A roof must be sloped enough to shed water or snow effectively, but as a roof becomes steeper it also becomes more difficult and dangerous to work on. We choose a 6:12 slope (i.e., 6" in rise for every 12" of run) for many of our projects. This 6:12 is a sensible compromise between competing concerns: a roof should be steep enough to shed precipitation, yet not so steep as to be unsafe (especially if non-professionals are involved). A steeper 8:12 slope is also workable, and should be considered especially if you want an insulated attic space.

As for finish material, metal is the most advisable. A roof fabricated of galvanized steel is long lasting, provides durable protection in all conditions, is readily available in most areas, and, when you consider the life-cycle cost of the building, provides good value for the life of your home. Asphalt shingles of high quality are, if carefully installed, a less costly alternative. With asphalt roofing, however, you do not have the possibility of recycling the roof at the end of its life cycle, and your ability to harvest rainwater is diminished. Landfills are full of asphalt roofing, but old metal roofs may find new life in future products.

Red Feather also makes use of roofing with continuous ridge and eave vents, which eliminates the need for attic vents in the gable ends of the building. This means you will have fewer openings around which to lath and stucco, and will minimize the chance of leakage in the critical situation where you have gables above bale walls. Refer to regional

Steep roofs require more material and are harder to work on, but the tradeoff is loft space.

Sheathe trusses with plywood, but stagger the sheets to provide a strong roof surface.

Covering a new roof with waterproof membrane

venting codes to be sure that your venting area is sufficient. With a good metal roof in place, you hopefully will not need to worry about roofing for over fifty years.

The materials you add to your structure in this chapter add weight to the RBA and serve to compress the bale wall. When the wall is fully loaded with the combined weight of trusses, framing, sheathing, drywall, trim, and insulation, leave the walls overnight to compress as fully as possible.

The next morning, nail metal coil straps (produced by Simpson Strong-Tie or similar) to the exterior sides of the RBA and the base plates. Bolt the bottom end of each strap directly to the face of the foundation. On each side of the corner of your RBA, attach a metal coil strap and run it diagonally down to the base plate making sure it does not come closer than 12" to a window or door opening. Coil straps should be placed at other points along the wall as per the design recommendations of a structural engineer. The straps are nailed through factory-punched holes with fasteners per the manufacturer's recommendation. These coil straps are very important, for they provide the direct structural connection from the RBA to the foundation. These straps are a defense against winds which could blow your roof off, and supply the major resistance your house will have to large uplift forces associated with major storms and earthquakes.

Coil Straps form x-bracing to resist wind and earthquakes.

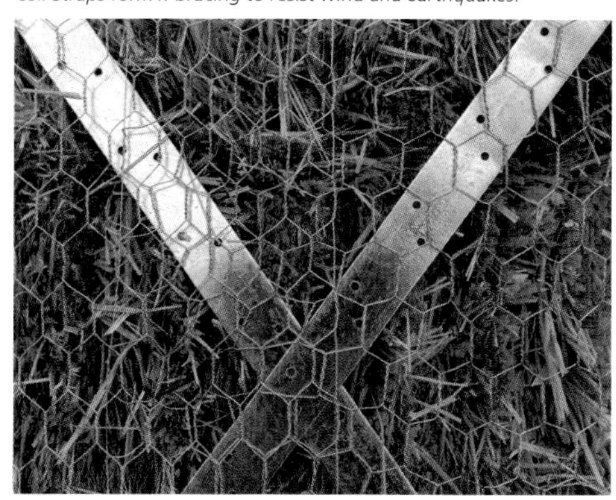

A metal roof protects bale walls for years to come.

Install ceiling drywall before starting stucco work to help compress the bale walls.

Straw bale walls go up quickly, and the experience is satisfying. Lathing, however, is time-intensive and tedious in comparison. In order to give stucco strength, especially for load-bearing straw bale structures, good lath work is vital. Thus, before you install lath, make sure that any wood you will be covering with lath is first sheathed with a layer of house wrap. Place this protective layer under the lath so it can protect wood items like window bucks from the water involved in applying stucco. You must also provide flashing at this point in order to channel moisture away from the interior of the wall.

FLASHING AND WEEP SCREED

LATH

BEFORE ATTACHING LATH

ATTACHING LATH

METHODS OF ATTACHING LATH

TYING SHARP ENDS

LATH AT CORNERS

LATH AT ELECTRICAL BOXES AND SWITCHES

LATH AT WINDOW AND DOOR OPENINGS

Flashing and Weep Screed

Begin by attaching a weep screed around the base of the building's foundation, and metal flashing above all window and door openings. Red Feather uses commercially available channel shapes, and fastens them directly to the base plates where they meet the foundation. Weep screed is manufactured with weep holes in the screed trough that permit water to exit the bale wall. Weep screed also provides a convenient reference for cutting lath sheets. If you fasten these sheets so they terminate in the channel portion of the screed, then their sharp ends will terminate in a screed channel and will be out of the way of future trowel work.

Lath

You must completely cover a bale wall with galvanized lath if the wall surface is to hold stucco. Red Feather uses expanded metal lath (also known as diamond lath) and common 20 gauge chicken wire. Diamond lath is used in areas that need more support, or will receive more wear—such as corners and window and door pockets—and horizontal ceiling areas like those above the windows that you will finish with stucco.

Note: Diamond lath is fabricated with a specific orientation to prevent stucco from falling off the wall while it is still wet. Use care in cutting and placing diamond lath sections so that the small diamonds act as cups and hold the liquid stucco against the bale wall rather than dumping it out on the floor. Close inspection of the diamond lath material will reveal this material's preferred orientation.

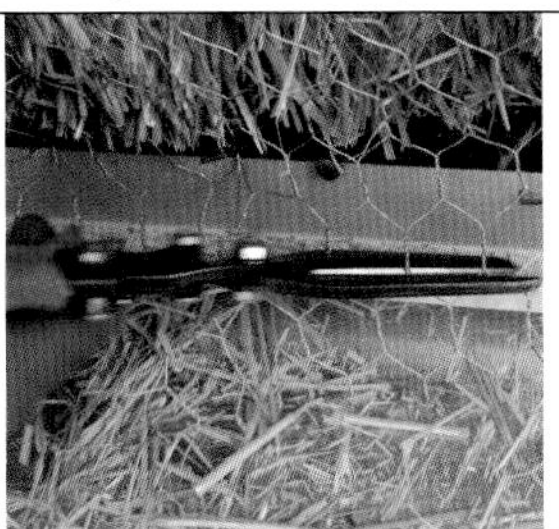
Trimming lath within the weep screed

Cover flat portions of bale walls inside and out with standard chicken wire.

Use loose straw to fill the gaps.

Before Attaching Lath

Before wrapping the bale walls in lath be sure to cover any wood portions of the structure—toe-up plates, door and window bucks, the RBA—with house wrap (such as Typar). Ideally this wrap should completely cover the wooden elements and be stapled between the buck and the bale wall. This layer protects wood assemblies from the stucco's moisture. House wrap, however, does not hold stucco. Therefore, apply diamond lath, not chicken wire, over all protected wood surfaces.

Before covering the walls with lath, it is also a good idea to stuff straw in any remaining gaps or cracks between bales. Some straw stuffing will be necessary to fill gaps in the wall, around windows and doors, and especially at the connection points between RBA and walls and between floor plate and walls. The object is to fill empty gaps with straw—not with stucco that will be applied later.

Attaching Lath

You must do a good job of putting lath up inside and out if you want your stucco application to progress smoothly. Your goal is to end up with a uniform layer of lath around the entire straw bale portion of your building without leaving any sharp edges or bits of wire sticking out that could catch a stucco trowel.

It is best to begin lath application on the flat field portions of your bale walls. Cover these fields with vertically-run sheets of chicken wire that overlap by about eight inches. At the top and bottom of the wall, fasten the chicken wire to the base plate and to the face of the RBA with pneumatic staples. Red Feather typically orders sheets of diamond lath that are at least as long as the finished ceiling height, and pre-cuts wall-height sections of chicken wire from a roll so that we also have a ready supply of sheets on hand.

When attaching chicken wire sheets near wall corners, be mindful of the width of the diamond lath sheets you have stocked, because your next step will be to fasten sheets of diamond lath around door, window, and corner areas. Ideally leave about a foot of bare bale wall exposed between the edge of the chicken wire and the wall corner on each side. Then attach a single sheet of diamond lath, folded lengthwise, in the corner to provide a base for stucco. This sheet must overlap the chicken wire sheets adjacent to it. This practice saves lath material and time because it creates the fewest inter-sheet seams, and joins finished edges which are easier to stitch together.

If you stretch the chicken wire and diamond lath sheets tightly when stapling them to the base plates and RBA surfaces, your work will go more quickly since tight sheets require less tuning later.

Buck exteriors are covered with house wrap. Note that each piece laps to shed water.

You can join lath in many different ways, including twisting sheets together with a pair of pliers.

All bucks must be wrapped prior to stucco application.

U-shaped landscape ties fasten lath to a bale wall.

Methods of Attaching Lath

To fasten lath sheets to straw bales, use galvanized landscape pins. Use galvanized pneumatic staples, however, to fasten lath sheets to wood components such as plates, RBA surfaces, and window and door bucks. The best fasteners for attaching lath to wood are galvanized roofing nails for they are designed to seal themselves to a waterproof membrane and thus make fewer holes which could admit water. Staples are typically faster than roofing nails, however, but whatever fasteners you use, be sure not to puncture the waterproofing membranes any more than necessary during lath installation.

In the interest of exercising some quality control during lath work, it is a good idea to have someone with experience insure that the lath is being properly installed. A tight fit to the bales is preferred. Ideally, the lath should not be able to be pulled away from the bale. Take care, though, to avoid over-tightening since this may unwind the lath itself or pull it from its point of attachment.

Tying Sharp Ends

The process of tying sheets of lath together—both chicken wire and diamond lath—is comparable to making a quilt out of different pieces of fabric. And, as in quilt work, this "tying together" is the most onerous part of lath work.

There are many ways to sew lath together. A paint-can opener, a screwdriver, a good set of pliers, and tough gloves are some of the tools Red Feather uses. The trick is to bend and twist enough individual wires from the edge of one lath sheet over in such a way that the sheet becomes firmly attached—sewn—to its neighbor sheet. In terms of finish, the main objective is to turn the sharp ends of all wires used for joining sheets back toward the bale wall so as not to create trowel-catching obstacles. Landscape pins are used about every 16" on seams and elsewhere as needed.

A given sheet of lath will either have a site-cut rough edge, with plenty of loose wire ends that can be used to tie lath sheets together; or a factory-finished edge with no such ends. Therefore there are essentially three edge conditions to consider when joining two sheets of lath: 1) joining a finished edge lath sheet to a rough edged lath sheet, 2) joining two rough edged sheets together, and 3) joining two finished edges together. In the first two cases, when you have at least one lath sheet with loose edge wires available for sewing, you can join the two sheets using these wires and the tools mentioned above.

When you are joining two sheets of finished-edge lath, another method may be employed. You can either make use of lengths of wire from a spool to link the sheets together (it is also a good idea to have a spool of baling wire on hand for areas where you will need longer pieces of thread, or you may cut into one of the finished edges to provide wire ends as required to sew the lath sheets together. For example, in order to fasten one finished-edge sheet of chicken wire to another finished-edged sheet, you can use wire cutters to snip into the edge of the chicken wire every six inches or so, and use the resulting loose ends to tie the sheet to its neighbor. You should also use landscape pins on such seams to help with the overall task of bending sharp ends back into the mass of the bale wall as well. In order to tighten a panel of lath that is too loose, take a pair of needle-nose pliers and go along twisting the slack out of the lath field at regular intervals.

Lath at Corners

At interior and exterior corners you should apply full-width sheets of diamond lath. If care has been taken to attach chicken wire so that a sheet of diamond lath will overlap the chicken wire on each side of the corner, you will save yourself a great deal of patching and splicing. Do not fold these corner sheets if a rounded corner is desired.

After securing the corner lath strip with staples at the top and bottom of the wall, sew and pin each vertical edge of the diamond lath corner sheet to the chicken wire sheets already on the walls. Then proceed with tightening the corner lath areas as you did on the walls.

Lath at Electrical Outlet Boxes and Switches

Another important use of diamond lath is to cover areas around electrical switches, plugs, and other fixtures that you will flush-mount in the bale wall. In such cases, small rectangles of diamond lath are cut 2–3" larger than a given electrical box in both directions. With a marker trace the box onto the lath and then mark an X from the corners of the box. By cutting on the lines of the X down to the size of the box, and then folding back the resulting triangles of lath, an open-sided box of bent lath can be made to fit snugly around an electrical outlet box. With its lath surround, an electrical box can then be set in a socket in the wall, landscape pinned in place, and then tied into the field lath of the bale wall.

Lath at Window and Door Openings

You must devote a great deal of attention and time to providing good lath coverage around your window and door openings. Since the buck assemblies for these openings are constructed of plywood, they must be protected from wet stucco. Once house wrap has been affixed over all wood areas, lath can be added. The process is as follows:

Apply house wrap and/or waterproof membrane to all interior buck surfaces, extending it inward so as to cover the edge of the plywood and, outward fastening it between the exterior of the buck and the bale wall itself.

Cut a sheet of chicken wire to the size of the opening where you are working. Wrap this sheet over the corner formed by the bare straw bale wall and the inside plywood face of a window or door buck and attach it with staples. Beginning inside the buck, secure and

Use pre-formed lath corner strips (shown here) or diamond lath at inside corners.

Use a rectangular piece of diamond lath around electrical outlets and switches.

shape the corner of the bale wall and the loose straw insulation you have added between the buck assembly and the bottom of the RBA. Stuff loose straw into any gaps while working to form an even contour at the corner joining the bale wall and the inside of the opening. By creating an even curve of lath at the edges of all openings, and strengthening the resulting soft edges by stuffing them full with straw, you will be able to provide visually pleasing corners at openings that are strong enough to hold stucco. Finally, landscape pin and tie the loose edges of the lath back into the bale wall.

To make an even contour for stucco below a window, staple a short piece of lath from the sill to the base plate. Place loose straw behind this lath, shape it, and then staple it to the window buck.

Cut and have on hand 3/4" plywood screed strips about 3–4" wide (cut from scraps whenever possible). These can be ripped out of plywood stock of various lengths on a table saw.

Secure a continuous band of screed boards around the frame on the interior of the buck opening closest to the window frame. These boards will provide a reference for the depth of stucco to be applied in the opening—the same 3/4" thickness as the plywood screeds—and will give an edge where you can end the stucco application before reaching the door or window frame.

Install curved strips of diamond lath to create even contours at the corners of windows and door openings. Begin this process with a pre-cut piece of diamond lath long enough to reach and overlap the lath on the adjacent wall surface. Butt one end of the diamond lath up to

Place straw behind lower lath sheet, shape it, and then staple it to the window buck.

Gloves, needle-nosed pliers, and patience provide a good layer of lath around a window.

This doorway is ready for stucco.

the window frame (you will need to fold it slightly to get it back into the corner of the buck assembly) and staple it to the interior of the buck; then splay out the other end and staple it to the RBA where possible and then sew it to the chicken wire on the bale wall. The flexible nature of the diamond lath allows you to make clean curved transitions where the corners of the window head (or sill) meet the walls.

Apply another layer of chicken wire from the inside top of the buck (near the top of the window frame) back to the RBA. This provides a double layer of chicken wire over the waterproof membrane protecting the top of the buck. Then apply diamond lath to all interior faces of the buck and stretch it around to meet the bale wall where you can landscape pin and sew it into the lath field. Once the layers of chicken wire have been covered with diamond lath, they will provide space for the stucco to engage in the lath. This process provides better adhesion for the stucco on the overhanging flat surface above openings.

Note: Work for consistency—especially with respect to the radius of the curves at the corners of each window and door pocket—for an attractive appearance throughout your building.

Left and right: Special attention must be given to the lath forming the soft curves at interior window corners.

A well-lathed window: diamond lath overlaps chicken wire panels at the corners, and waterproof membrane covers the entire wood window buck.

Note: Triangular plywood gussets at top hold the buck square until after stucco completion.

Stucco application is a wonderful group activity that can be shared and enjoyed by elders and kids alike. With an enthusiastic crew, stucco application is quick and satisfying, producing a beautiful textured wall that will protect the bale wall from pests, moisture, and fire while adding structural strength. Since labor typically costs more than materials for housing built in the United States, a group of people working together to build, lath, and finally stucco a straw bale wall can significantly lower the cost of a home.

INSTALLING STUCCO SCREEDS

MIXING STUCCO

STUCCO SAFETY NOTES

STEPS FOR MIXING STUCCO

STUCCO MIX RECIPES

SCRATCH AND BROWN COATS

EXTERIOR FINISH COAT

INTERIOR FINISH COAT

STUCCO APPLICATION

THREE-COAT STUCCO

Installing Stucco Screeds

The first step is to install screeds—a screed is a piece of material that provides a guide, like a small dam, which tells you how deep you have applied stucco when you can no longer see the lath substrate. On a building's exterior, Red Feather fastens commercially available metal screed products to the RBA and to the base plates to define the edge of stucco coverage at the top and bottom of the wall. Within a building, Red Feather uses strips of plywood ripped from scraps as screed boards. Your objective is to install these plywood screed strips wherever a stucco wall shares an edge with a finished, flat surface. The screeds are applied flat, using either screws or pneumatic staples, with one edge pressed to the flat surface of the floor, ceiling, or a buck frame depending on the situation.

Note: The top interior edge of the base plate, or simply a piece of J-mold held 1/8" off the finish floor, (described in Chapter 3) acts as the screed strip for the base of your interior walls.

At the ceiling a continuous line of screed strips is fastened to the top of the RBA to transition between the stucco wall and the finished drywall ceiling. This 3" or 4" band of plywood screed around the top of the interior walls will provide an upper boundary to terminate the stucco wall coats, and a margin of protection against the wet stucco damaging the ceiling drywall installation. Also use screed boards ripped from scrap plywood to frame the inside of window and door bucks. Cut these to the desired thickness of the stucco and apply them to inside corners where the 2x4 window frame and the plywood skirt of the bucks meet. Make sure to push these screed boards up flush to the window frame; they will keep stucco from touching the clean edges of the window and door assemblies. In both cases, screed assemblies serve as an attachment point for interior trim, whether it be baseboard, cornice trim at the ceiling, or finish trim around door and window frames.

Note: An experienced design professional should detail and inspect screeds and flashing at window and door rough openings. Be sure that experienced professionals install metal exterior screeds. For example, the weep screed at the bottom of exterior walls must lap over the top of the foundation wall by at least one inch in shingled, weather-shedding fashion. Flashing products only keep water out of the building when they are properly installed.

Screed strips limit stucco, protect ceiling surfaces, and provide a continuous nailing strip for trim.

Mixing Stucco

The first task is to prepare a mixture of wet stucco. You will need a specialized group of tools, and a few hard working helpers to do the mixing. Tools for this step are described in some detail because conventional stucco mixing and application techniques must be slightly modified for straw bale construction. Also, in conventional construction, stucco is often sprayed onto wall surfaces. This may be considered for straw bale projects, but Red Feather uses large work groups to apply it by hand. We feel that this method is safer for less-experienced workers, and it reduces overall costs for volunteer-based projects.

To prepare stucco, ensure that an experienced person—someone fully knowledgeable about straw bale stucco application—is available to lead the stucco team. This lead mixer should have a team of at least four people. Three will shovel, measure, and prepare materials (sand, cement, lime, etc.) at the mixing site, and may help to move mixed stucco to the project walls in wheelbarrows. One will oversee the quality and quantity of the stucco being used on the walls, and communicate stucco demand and quality control issues back to the mixing team. Any additional workers should be added to those on the wheelbarrow detail.

Locate mixing equipment and supplies on level ground free of overhead obstructions in close proximity to the building project. Organize this work area to facilitate the free flow of wheelbarrow traffic. Most importantly, create an unobstructed path for full wheelbarrows—often weighing as much as 90–100 pounds—to move between mixer and building. Also make sure the incoming-materials path does not cross the outgoing wheelbarrow path. Two team members should work together to empty the heavy bags of cement into the mixer. It is best to create a short platform on which to rest the bags while one pours their contents into the mixing drum. This will prevent waste, and will help keep the mix accurate.

Stucco Safety Notes

- Everyone handling powdered cement, sand, and lime must wear protective gloves and eyewear, and a filtration mask.
- Water must be kept close at hand in case a worker gets chemical contact burns from Portland cement and lime. Also encourage workers to rinse areas of their body that have come into contact with wet stucco mix frequently to prevent burns.
- Stucco mixing is hard work, so it is important to allow for frequent breaks.
- Be sure to rotate work details among team members often to increase safety and decrease fatigue.

A stucco team and its mixer

Lots of help makes stucco work go quickly.

Steps for Mixing Stucco

1. Select an appropriate site for the mixing station.
2. Assign team member duties and review stucco safety issues.
3. Post a copy of the stucco recipe at the mixing station.
4. Review stucco process and recipe with work team.
5. Check the mixer engine for gas and oil.
6. Check site for proper location of materials and pattern of work flow.
7. Attach water hose and gather necessary wheelbarrows, buckets, and shovels.
8. Fill water buckets to proper levels as indicated by the recipe.
9. Start mixer.
10. Add required amount of water.
11. Add about half of the sand.
12. Add Portland cement.
13. Add lime.
14. Add remainder of sand.
15. If necessary, add additional water until desired consistency is met.
16. Add fibers (if used) per manufacturer's instructions.
17. Mix 3–5 minutes after all ingredients have been added.
18. Dump mixed stucco into wheelbarrows and transport it to the building site.
19. Add water immediately after you empty it. This prevents stucco from drying out on the inside before you make the next load. Use a rubber hammer on the outside of the mixer drum to loosen clumps of stucco stuck to the inside.
20. Repeat the entire process. Help workers anticipate material and water needs for subsequent batches.

Note: There will be times when you will need to adjust the quantity of water, fiber, sand, and lime in order to produce a "wetter" or "dryer" mix. The mix supervisor will be able to see how fast or slow the application is going and what type of mix will be needed to keep the stucco work process smooth.

Stucco Mix Recipes

In practice a batch of stucco from a typical mixer covers approximately 80 square feet for a scratch coat, and a somewhat larger area for brown and finish coats (see below). It is helpful to measure materials by volume. To do this you can mark containers with fill lines. It may also be helpful to translate the part ratios listed below into other measures: bags of Portland cement, shovels of sand, 5 gallon buckets of water, and so forth.

Scratch and Brown Coats

Mix:
¼ (to a maximum of ½) part type-S lime
1 part Portland cement
3–4 parts masonry sand
Water as required to achieve workability—that is, a consistently moist mixture that will keep its shape on a trowel.
Optional: fibers per manufacturer's recommendations

Note: If it is necessary to adjust the mix for workability, add water, sand, or cement. Do not add more lime than listed above. Use as little water as required for a workable mix. Water should be minimized because excessive water decreases the strength of the finished stucco, and may cause cracking.

Exterior Finish Coat

Mix:
1–2 parts type-S lime
½ part White Portland cement
½ part Portland cement
3 parts silica sand (grade 20)
Water as required to achieve workability
Color admixture per manufacturer's instructions

Note: Check with the Portland cement manufacturer to be sure that the White Portland and Portland cements may be mixed together. (This also applies to the interior finish coat described below.)

Interior Finish Coat

Mix:
1–86-872 parts of type-S lime
1 part White Portland cement
3 parts silica sand (grade 20)
Water as required to achieve workability
Color admixture per manufacturer's instructions

Note: Optional finish coatings: Red Feather has recently been using several quality products to better inhibit water penetration while allowing vapor permeability. If the three coat stucco process described above is used, painting a layer of siloxane over the final coat does much to reduce water transmission. Another option is to add a final elastomeric coat that can be painted or rolled on in place of the third coat of stucco to provide a flexible, waterproof, yet vapor-permeable finish surface in a range of colors. Both products can be sourced through the Sika Corporation at www.sikacorp.com.

Stucco Application

During stucco application, Red Feather uses straw bales in a variety of ways. It is very useful to set pieces of scrap plywood on top of straw bales or saw horses to make improvised tables where stucco, unloaded from the wheelbarrows, will be easily accessible to those applying it to the bale walls. However, stucco awaiting use on the tables may need to be moistened (tempered) to keep it workable.

Note: Do not temper the mix more than once. Also do not apply any stucco that has been out of the mixer for more than 1½ hours; such stucco should be discarded.

When applying stucco to the exterior walls, straw bale scaffolding will prove very useful, so it is best to leave portions of the scaffolding in place after you complete your lath work. While applying interior stucco, it is helpful to place loose straw at the base of the walls to catch falling bits of stucco. Remove this straw periodically to keep the floors dry and clean.

A hock and trowel are the primary tools you will use to apply stucco, and rounded trowels (also known as "pool" trowels for their typical use in swimming pool construction) are better suited to straw bale stuccowork than trowels with sharp corners. Pool trowels are able to create the gradual contours and corners natural to straw bale buildings. Similarly, sponge trowels are better than hard trowels for smoothing brown and finish coats, since they are able to handle the surface irregularities of bale walls.

To apply stucco on a wall surface, start with a hock in one hand, and a trowel in the other. From the stucco table scoop up a portion of stucco with your trowel and drop it on the hock. Approach the wall, tilting your hock forward as you move in. Push the hock firmly against

The amount of sand in a mix is important. Be diligent to keep it consistent in each batch.

People of every age find stucco application satisfying.

Use essentially the same materials, tools, and techniques to apply stucco inside and out.

Preparing to stucco a well-lathed window. Note the diamond lath around opening.

This window shows the pleasing organic forms typical of straw bale walls.

the wall in an upward manner, pressing stucco firmly through the lath layer and into the straw bales (or into the scratch-coat grooves, if you are applying a brown coat). Now use your trowel to smooth out irregularities and insure a continuous surface. Do not spend too much time fiddling with the stucco, that will come later; at this point the object is to get stucco on the wall.

Work from the bottom of the wall to the screed at the top. Then, using your trowel in big sweeping curves, smooth the stucco evenly. You will have to practice with the tilt of your trowel to avoid digging in, but you will soon find the correct motion.

This is the most efficient and productive way to apply stucco, but truthfully there are many workable techniques. Stucco is forgiving and you have a fair amount of time to correct mistakes while the stucco remains wet and pliable. Do your best to get the stucco pushed well into the wall surface and try to limit material waste. You will find your stucco rhythm with a bit of practice.

Three-Coat Stucco

Stucco is typically applied in three coats: a scratch coat, a brown coat, and a finish coat. The first, or scratch, coat covers straw and lath; then you scratch this first coat with a metal comb to make small grooves—known as keys—parallel to the floor so that a second, or brown, coat will adhere to it. Later, you will apply a finish coat over this brown coat.

Apply the first, or scratch coat directly to a lathed bale wall. This coat should be at least 3/8" thick, but it will probably be a bit thicker on irregular parts of the wall. The object here is to get some of the stucco into the bales, and to cover the lath layer. Complete coverage is necessary to provide a good bond between lath and straw. It is crucial that you apply enough stucco to the lath so you can make approximately 1/8"-inch horizontal scratches (keys) with a scratch comb. These grooves provide a surface for the next layer of stucco—the brown coat.

Note: The scratch coat should be kept moist until the brown coat is applied.

Apply a brown coat as soon as the scratch coat is strong enough to support it. This can be done soon after the scratch coat is applied, but it must be finished within twenty-four hours. Again, be sure the first coat is kept moist. Ideally scratch and brown coats will cure together and behave as one layer of stucco, because simultaneous curing creates a stronger bond, and results in a stucco coating that will resist cracking and water infiltration better. In hot or dry weather, you will probably need to mist the scratch coat in addition to covering and shading it with tarps.

Note: Stucco should not be applied if you anticipate temperatures below 40 degrees. This is good news for stucco workers, since wet work in such weather is unpleasant even if it were advisable.

Groups make stuccowork go quickly. Note screed board at the top of the wall.

Scratched wet stucco provides a base for a brown coat.

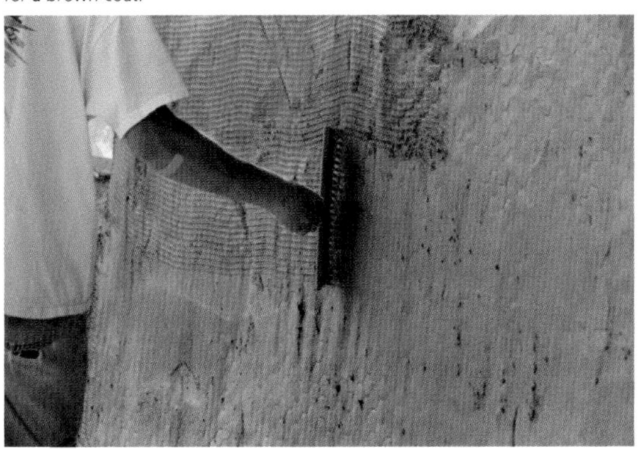

A water sprayer facilitates an even curing process.

After making sure the scratch coat is moist, apply a brown coat—a stucco layer of about 3/8" thickness. The brown coat needs to be smoothed flat along the outer edges of the screeds. The object here is to avoid bumps in the surface where you will be attaching flat pieces of trim or wood casing.

After application, you must keep the brown coat moist for a period of at least two full days. In hot and dry conditions, mist stucco walls once in the morning and once in the late afternoon, and use tarps to protect stucco surfaces from evaporation. These precautions will provide the walls with the moisture they need to cure properly.

Ideally a finish coat is not applied until several months after the brown coat has cured. But, if necessary, it may be applied as early as eight days after the scratch and brown coats have cured. It is best, however, to leave several months between brown and finish coats. You will want cracks (if any) to occur before you apply your finish coat because it will fill and seal them. Apply the finish coat to a thickness of at least 1/8", and then use a trowel to smooth it out in a consistent pattern. Cure this finish coat for two full days or longer.

The three-coat stucco process described above results in a stucco surface that is at least 7/8" thick. And although Red Feather avoids building designs that feature extensive stucco on horizontal surfaces, the flat portions above window openings are an obvious exception. For these horizontal (or ceiling) surfaces, apply stucco in the thinner coats. Red Feather recommends 1/4" scratch and brown coats, and a 1/8" finish coat, for a total thickness of 5/8".

Note: Give the first coat on horizontal surfaces a bit more time to cure before adding a brown coat. Stucco in horizontal locations needs to be more solid in order to stay in place.

The most difficult places to stucco are typically above window openings. Careful lath work and patience are crucial.

Working the scratch coat toward an exterior corner

Rinse all stucco tools after use.

After you have given your straw bale walls their stucco coatings, it is time to set to work finishing the inside of your home. The perimeter bale walls of a straw bale house are typically covered with stucco, but the interior walls are often made of less bulky materials. Since the interior walls are not structural they may be located nearly anywhere. Material choice for interior walls is another personal decision. Will you build with conventional drywall or use one of the many agricultural byproduct materials now available?

INTERIOR WALL FRAMING

OTHER MATERIALS OPTIONS

NOTICE
KELLER

Note: This chapter is not intended to describe everything one must know to frame the interior of a house. It is rather an account of how conventional light-frame construction interfaces with straw bale walls, and it assumes that a reader has some knowledge of conventional construction, or will seek out other resources on this subject.

Interior Wall Framing

To partition interior spaces, construct stud-framed interior walls on the interior plates you will have laid out on the foundation. We recommend placing most of the electrical work and essentially all of the plumbing work within the interior walls. Keeping these systems out of the exterior bale walls facilitates change; it also reduces the likelihood of water leakage or fires within your straw bale walls.

You can use conventional 2x4 studs to build interior partitions. Try to select Forest Stewardship Council (FSC) certified lumber if it is available. Red Feather typically uses drywall over typical wood framing for most of the interior walls, but there are also good alternatives. Drywall has the advantage of being widely available, fire-resistant, and relatively benign from an environmental perspective. Stucco may also be applied over lath on a stud wall frame. Several straw bale homes have been built with conventionally framed interior walls that were subsequently stuffed with straw, lathed and coated with stucco to give a surface finish more in keeping with the aesthetic of the straw bale exterior walls.

Where a 2x4 framed wall meets the bale wall a solid connection needs to be made. The RBA and base plate components of the load-bearing straw bale wall system provide good surfaces for tight connections. Simply fasten the end stud of a conventionally constructed wall section to the top and bottom of the wall. If you desire a positive connection along the full height of the wall, you can key a 2x4 into the bale wall before applying lath and stucco in order to provide a continuous nailing surface.

Interior walls are easy if floor and ceiling are flat and parallel.

Other Materials Options

Straw products: Straw bales are typically too thick to be used for interior walls; however, a bale wall between your building and a future addition would give you a handsome deep doorway, privacy, and good thermal and acoustic separation. Other options for interior walls include agricultual byproduct panels on a stud-framed wall system and post-agricultural building systems that provide their own support. Compressed "ag-board" materials—made from compressed straw and/or seed hulls—make great companions for straw bale exterior walls (available from PrimeBoard Inc., North Dakota). These ag-boards are typically fastened to conventional stud-framed walls (made up of Forest Stewardship Council Certified lumber whenever possible). Self-supporting compressed-straw panel systems are available in several formats as well: a paper-faced compressed straw panel (Affordable Building Systems, Texas) makes good interior walls. Straw-insulated structural panels are also available for taller walls and walls that need more strength (Agriboard, Texas). There are also dimensionally stable compressed straw bale blocks to consider (O-blocks by Oryzatech, California). The leaner dimensions of these products make them ideal for interior partitions or for projects in regions where super-insulation is not required.

Adobe or Brick thermal-mass walls should also be considered, especially in colder climates. When brick, stone, or adobe interior walls surround a fireplace or stove—or when you locate thermal-mass interior walls to absorb direct sunlight—such walls work well to store the day's heat and stabilize indoor temperatures during cold winter nights.

Note: Professional electricians and plumbers should be engaged to perform all electrical and plumbing work. Any information provided in this handbook related to these systems is meant to assist professional tradespeople who may be unfamiliar with how their work is handled in the context of a straw bale building.

Opposite top: Drywall attached to the bottom of the trusses provides a finished ceiling.

Opposite bottom: Protected from the weather, one can take a well-earned rest.

Left: Straw bale walls contrast nicely with interior finishes.

The detailing possibilities for a straw bale building are nearly limitless. Design decisions concerning the shape of wood trim and moldings, the texture of stucco finishes, the color of paints and stains, and the inclusion of built-in straw bale features like benches, niches, and "truth windows" define the unique personality and beauty of each house. Decisions involving finish details are also an important way to involve less experienced builders in the construction process. At this stage of projects, Red Feather has also been able to enlist community artists in adding mosaics, murals, and wall relief sculptures that give character and beauty to a straw bale building.

WOOD TRIM AND MOLDINGS

WINDOW AND DOOR INSTALLATION

STUCCO TEXTURE AND COLOR

TRUTH WINDOWS

SITE WALLS

SITE WALL CONSTRUCTION

Wood Trim and Moldings

Red Feather typically uses a simple band of 1x4 cedar for trim. Cedar is attractive whether left natural or covered with paint or stain. Cedar also resists decay, so exterior applications hold up well. Red Feather installs cedar trim around door and window frames, and along the floor as a baseboard. This application of trim covers the joints between the flat surfaces of floor, window, and door assemblies and the organic planes of the straw bale wall. Trim may be omitted if you take care to do good lath and stucco work at every transition point in your building. If you take this option, you can install a metal screed molding (for example a J-mold) around window and door frames, and wherever else the bale wall meets the ceiling and floor. Also be on the lookout for salvaged wood or other items that you could use in the course of the project. Using salvage for trim and other applications lowers costs as well as demand on tree-based resources. Moreover, old elements in a building add to its character and aesthetic appeal.

For exterior trim a durable and weather resistant material like cedar is best. Exterior applications include: window and door trim, decks and patios, roof trim (at both fascia and soffit), and accent trim such as band and skirt-boards (i.e., trim at the top and bottom of the exterior wall). When adding exterior trim, keep gravity and the flow path of water in mind. Think where you do not want water to go—inside the bales, windows, and doors—and how you can "head it off at the pass." For example, Red Feather applies a 1x4 trim to the exterior of windows and doors. Your objective, here as elsewhere, is to create a trim design that sheds water. You also will want to install a drip cap (a commercially available metal flashing section) on top of the trim, and a sloped sill (or no sill) at the bottom of each exterior opening.

When installing exterior trim, apply a generous amount of quality sealant (one that will remain flexible through the wide range of temperatures your building will experience) to its back to insure a tight seal with the wall surface. You may find that you will need to scribe the back of the trim in the event it has a wavy surface. Attach exterior trim material with corrosion-resistant fasteners such as galvanized screws or nails. Follow the same steps to trim and seal around doors.

Window and Door Installation

Window installation (as opposed to the installation of window bucks, which you have already done) takes place after you have finished applying exterior stucco and color coats. Simply place the window in the opening, shim it slightly off the rough sill, fasten one corner to the buck and then level and plumb the window assembly. Fasten the rest of the window frame to the screed boards already in place within the window bucks. The window should end up flush with the building exterior and in this way eliminate exposed sills that could collect and trap water.

Install doors in a similar fashion. Bear in mind that you should make an effort to center window and door assemblies as perfectly as possible within the frame of the buck. This facilitates interior trim application, and improves the appearance of the windows and doors once they have been finished and trimmed.

Stucco Texture and Color

The choices you make in surfacing and coloring your stucco provide opportunities to personalize your straw bale walls. In a straw bale home, the entire wall surface literally becomes sculptural, and during the final stucco coat you can use different textures and colors to achieve a desired aesthetic effect. If you wish, you can add color to the final coat of stucco that you apply to the wall, as opposed to simply applying a surface coat of color to a stucco wall. This integrates the color into the wall, and makes the color more durable. A wide variety of stucco textures are possible for interior walls; on exterior walls, however, stucco texture should be as smooth (or as minimal) as possible in order to guard against moisture infiltration. A final (or third) color-impregnated elastomeric coating may be used over your stucco walls.

Truth Windows

A tradition in recent straw bale buildings is to build a "truth window" at a special place in the building's interior, since once your straw bale project is complete, it will be difficult to tell that the walls are actually composed of straw bales. Visitors see an attractive, thick-walled building, but they will most likely assume that you built it out of masonry or adobe as only stucco will be visible. In response to this, straw bale builders often incorporate a framed opening that shows a portion of the straw within a bale wall—a truth window. Such a window represents yet another way of personalizing your building, and telling the secret of its warmth, coziness, and structural composition. A truth window says: "This house is built of straw."

A Red Feather volunteer team and the house they helped build in three weeks. Note the cedar trim along the roofline.

A salvaged post finds new life in a porch assembly.

A truth window and its door

Another spirited truth window

Site Walls

A good way to make use of leftover straw bales is to construct site walls. Extra pieces of lath and treated lumber ends from plate framing can also find a home within a site wall—either alongside an existing building or a new straw bale structure. Site walls are built in much the same manner as bale house walls are, but can more easily take rounded and organic forms relating to site contours and conditions. These walls define outdoor spaces and can also be sited to protect the building from cold winter winds and driving rain. There are many design options for site walls, and since they only need to support their own weight, they are free to curve and change in both thickness and height to suit your taste and needs. Site walls can incorporate benches, decorative relief sculpture, and even arched portals. They are also a great way to get involved in straw bale construction before tackling an entire house, and a wonderful way to introduce kids to building with straw bales.

Site Wall Construction

Straw bale site walls need a high, dry foundation and a good "roof" just as a straw bale house does. However, it need not be as rigorously constructed as a house foundation; even a rubble trench or a dry-stacked stone foundation is appropriate for a site wall. Another option is to use urbanite—salvaged chunks of discarded concrete that can be used exclusively, or in combination with stone, taken from the site or from your foundation excavation. But whatever base you use for your site walls, it should rise above ground high enough to prevent the bales from any risk of exposure to surface water—at least six inches above grade.

Site walls are assembled in the same manner as straw bale walls (see Chapter 5). Bales are stacked in alternating courses and can be pinned with rebar, hardwood surveyor's stakes, or bamboo. Once you have built the desired formation, cover it with lath and stucco in much the same way you did your straw bale building. Keep the following guidelines in mind. First, before stuccoing, contour the top tier of bales in a form that will shed water well. Next, drape a layer of waterproof membrane over these topmost bales. Stuff some straw between the bales and the waterproof membrane layer to get a tight wall form to which stucco will securely adhere. Then, tie and landscape pin a layer of chicken wire to the entire bale wall surface (refer to Chapter 7 for more information on applying lath).

Opposite top: Straw site walls define outdoor spaces.

Opposite bottom: Wrap site walls with lath and stuff them with loose straw to create a rounded profile.

Right: Salvaged concrete blocks and treated lumber are one way to build a site wall foundation.

Once the lath work is complete, stucco should be applied quickly to lessen the chance of exposing the bales to the elements. Red Feather uses the same stucco mix on site walls as on a building's walls. Since routine plaster maintenance will be required for all stucco work—especially for the curved upper portions of site walls in severe climates—the use of the same stucco mix will make this work easier. Matts Myhrman has developed another good way of protecting the top surfaces of straw bale site walls. Using Spanish-style ceramic roofing tiles (which are available in lengths that will span a two-string bale laid on edge), he builds an attractive tile roof along an entire wall. You can give taller site walls—used for privacy—other types of conventional roofing hats as well, such as metal roofing or a range of materials. Just be sure you slope and detail your site wall roof sufficiently to prevent any accumulation of water.

Applying stucco to a site wall

APPENDICES

Advice and procedures focusing on safety appear throughout this book. Each chapter includes safety-related information that has been refined through Red Feather's years of project experience. While specific issues relating to safety are included alongside the work procedures and processes they relate to, the following are some more general notes about working safely on similar building projects. This is not a complete list of safety concerns, but rather a few safety reminders that Red Feather includes in the initial safety talks with new volunteers and community participants before building projects begin.

General safety and well-being reminders

- We recommend that building projects have at least one medical professional on site at all times.

- On summer builds it will often be very hot during the building process. Dehydration and other exposure-related issues are a very real concern. Please drink a lot of water before you feel thirsty to avoid such problems.

- In many areas of the country, trees, tall grasses, and poison ivy may be present on or near a building site. This vegetation may provide a happy home for ticks and other biting bugs. You must be mindful and vigilant to search your body for these insects.

- Thunder, lightning, rain, and hail should be expected. Winds can be severe at times, and we suggest that you bring adequate protection for such conditions.

- For several reasons, we must stress that smoking is not allowed on or near a project site. Loose straw is very flammable, and smoking on site threatens equipment, materials, and the structure itself. Likewise, everyone needs to be extremely careful when handling anything that might spark a fire.

- A clean site is a safe site. All building projects should have persons assigned to collect and properly store loose equipment and tools. Loose items can be tripping hazards. Power cords, rope, hoses, and other similar gear should be kept well clear of circulation paths.

- Be aware of people around you and obstructions above you. Always check your surroundings before carrying large items to and from the construction site.

- Be aware of yourself. Fatigue and exhaustion can lead to serious injury. Know your limitations and rest accordingly.

- Lift properly. Building materials and construction equipment are generally heavy. Bend at the knees when lifting, and know your personal limitations.

- Do not use a tool that you are unfamiliar with until you receive training. Power tools such as nail guns, saws, and mixers can cause serious injury or even death.

- Service tools regularly, and make sure they are in proper working condition before using them. Improperly maintained tools and equipment can become serious safety hazards.

- If you do not understand a procedure, a process, or any other situation on the building site do not be shy about asking questions of more experienced persons. Understanding the materials, tools, and processes involved in a construction project leads to safer practices and better work.

3D EXTERIOR VIEW - AERIAL
3D EXTERIOR VIEW - NORTHEAST
3D EXTERIOR VIEW - NORTHWEST
VIEW - SOUTHWEST
BUILDER'S SET
CONSTRUCTION DOCUMENTS
A0

APPENDIX B | **DESIGN DRAWINGS**

The following detail drawings are shown to clarify the connections and assemblies described in this book. They do not represent complete drawings for a construction project and should not be substituted for professionally produced construction documents tailored specifically to your project, site, and region.

Plan 1

A modest house with a bathroom, kitchen core, and approximately 900 square feet of interior space. The interior walls are easily configured for one bedroom. The house is shown here with a mud room, barrier-free ramp, and side porch.

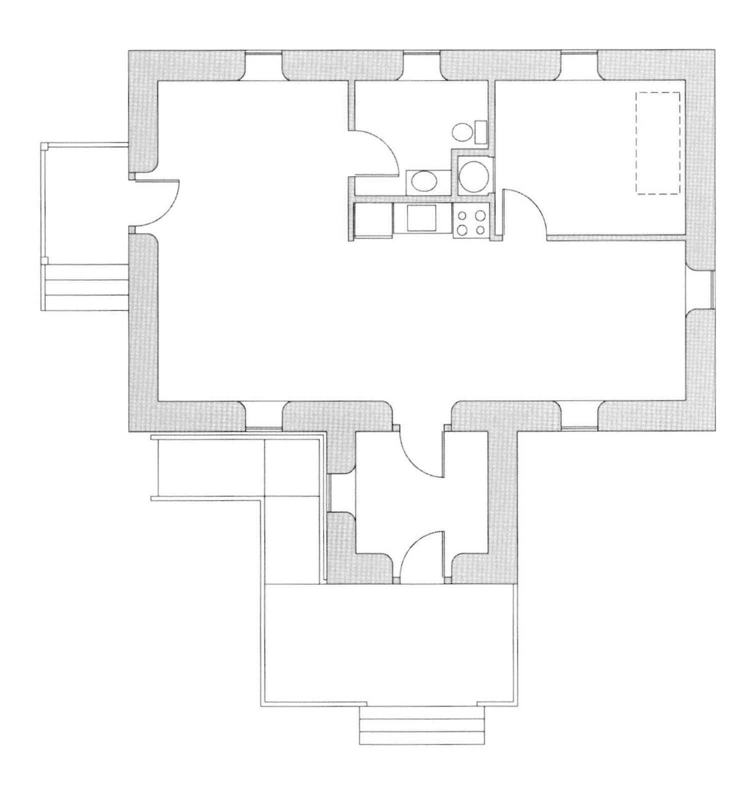

Plan 2

A somewhat larger two-bedroom house with a bathroom, kitchen core, and approximately 1,100 square feet of interior space. The interior walls define the bedrooms, utility room, and closets. The house is shown here with a mud room, a barrier-free ramp, and a side porch.

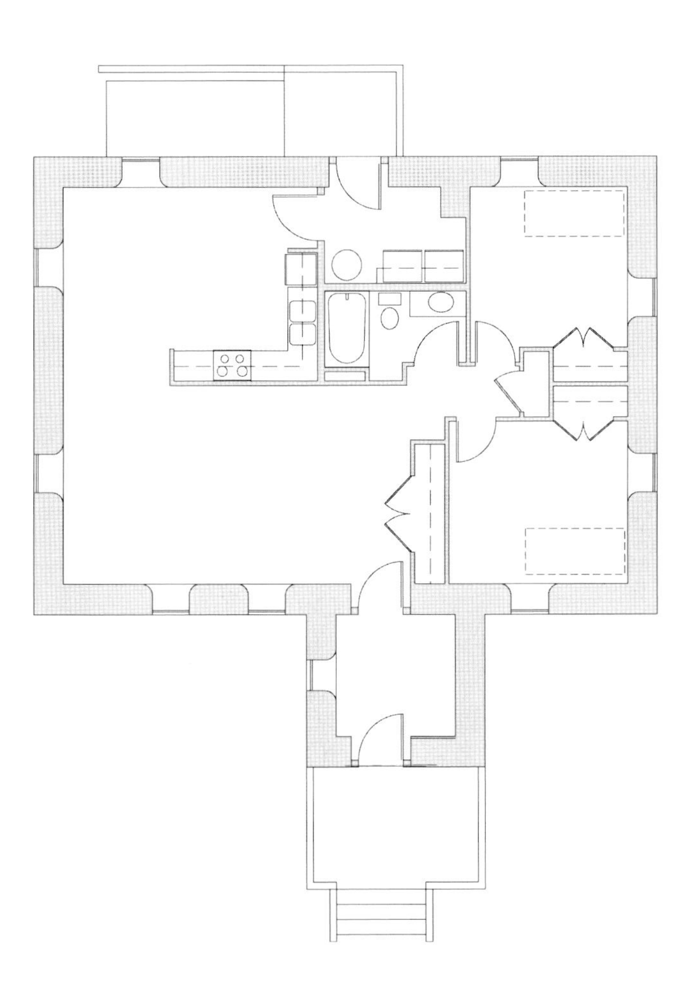

Plan 3

A larger two-bedroom house with a bathroom, utility, kitchen core, and approximately 1,500 square feet of interior space. The interior walls define the bedrooms and the utility room. The house is shown here with a mud room and barrier-free ramp. Note the interior bale-lined airlock, the straw bale buttresses that help partition the bedrooms, and the multi-purpose nooks that may be used for sleeping or storage.

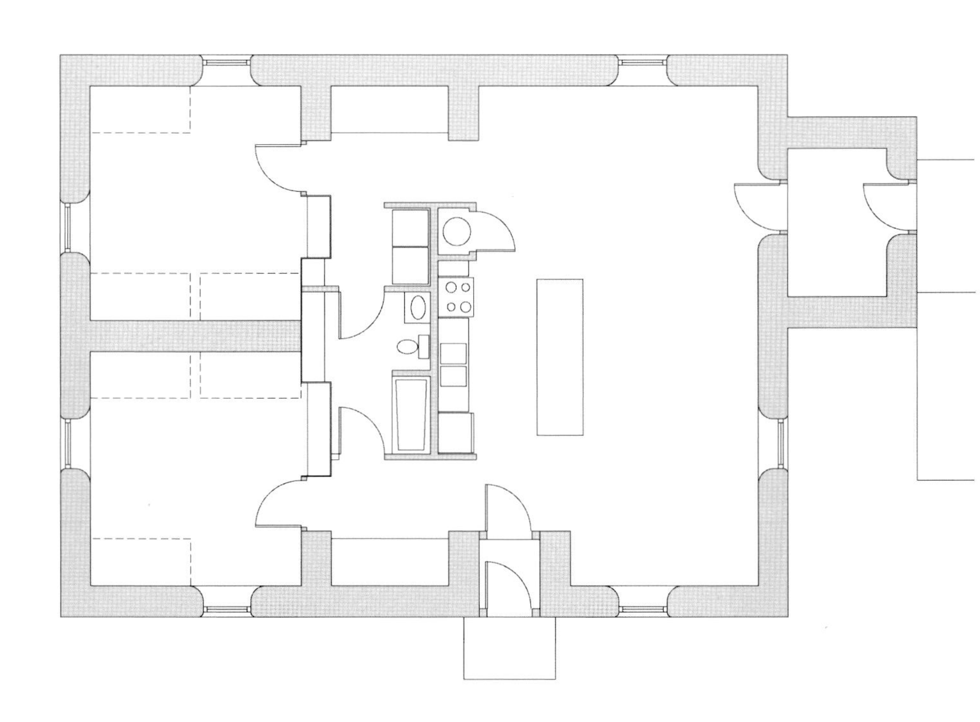

Plan 4

An efficient two-bedroom house with a bathroom, laundry, kitchen core, and approximately 1,050 square feet of interior space. The house is shown here with a barrier-free ramp. A mud room may be formed by the addition of a wall and door between the entry and the main space. This compact building core minimizes interior framing as well as water and electrical runs and provides a generous open living space with the possibility of a vaulted main room and a loft above the core and bedrooms. This is the plan used to illustrate the bale diagrams on page 123.

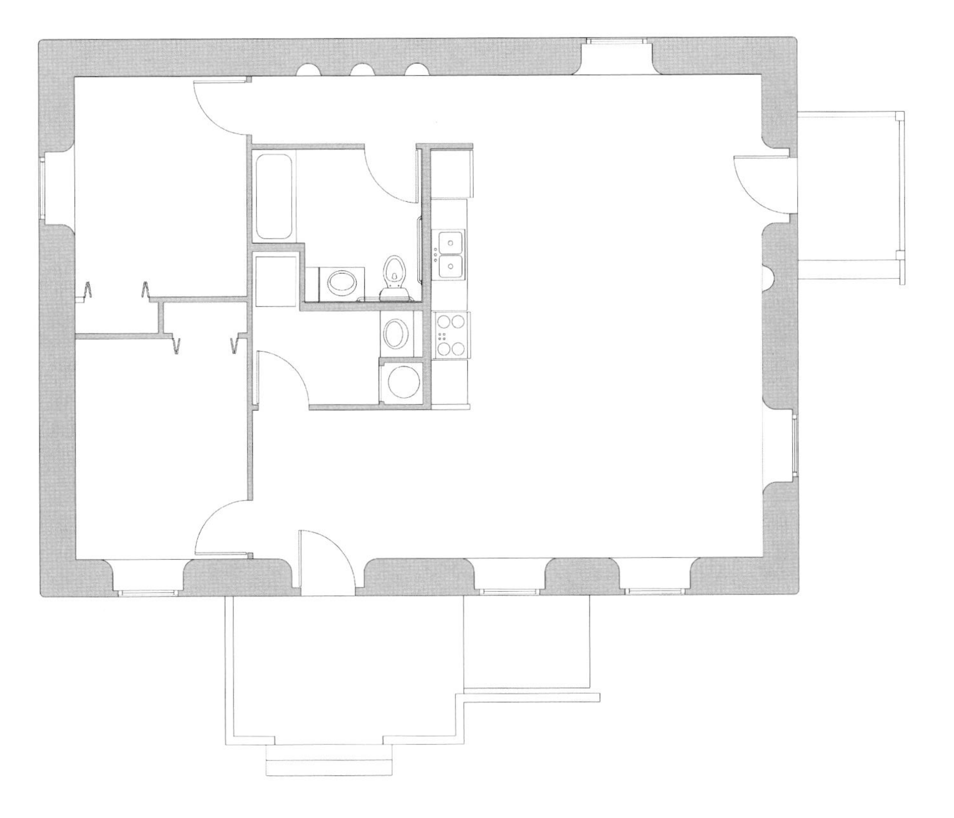

Bale Diagrams

This page: elevations of a simple straw bale building (see Plan 4). Facing page, bottom: elevation bale diagrams of the same building. Note: Loose straw infill closes the gaps above window and door openings.

Bale Diagrams in Plan

The plans shown here illustrate how bale diagrams are laid out. These plans relate to Plan 4 (presented earlier in this section and again here at the right) and correspond to the elevation bale diagrams on the facing page. The upper left plan shows the bale diagram for the odd numbered bale courses (the first, third, and fifth courses). The upper center plan shows the location of bales on the even numbered courses of the same building (the second, fourth, and sixth courses).

Drawing bale diagrams is useful in order to locate openings and to obtain the bale count for a specific design. Such diagrams should be planned to minimize the need of bale customizing on a project and should be present during the bale wall raising so that windows and doors are located properly and work flow is assured.

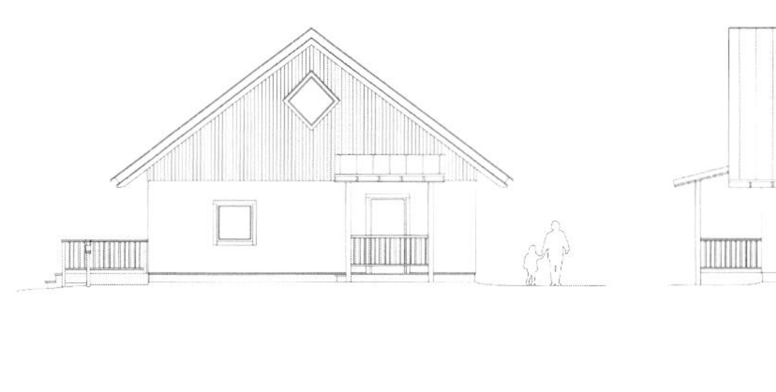

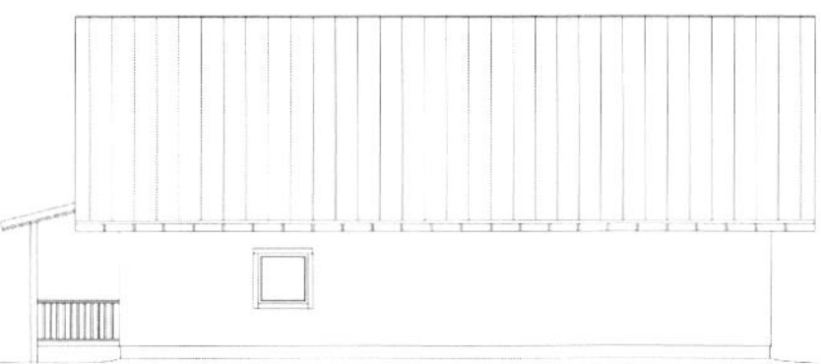

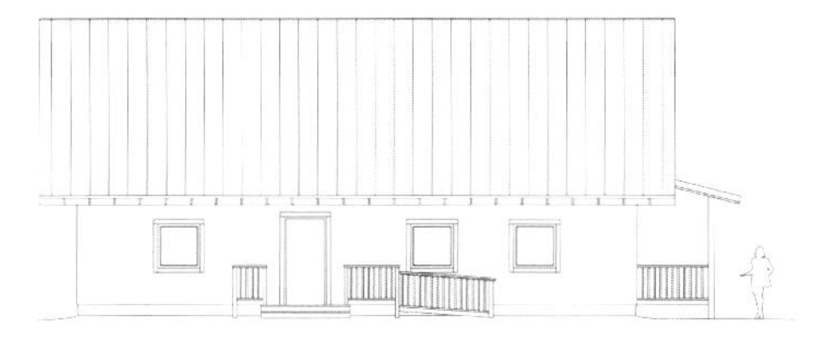

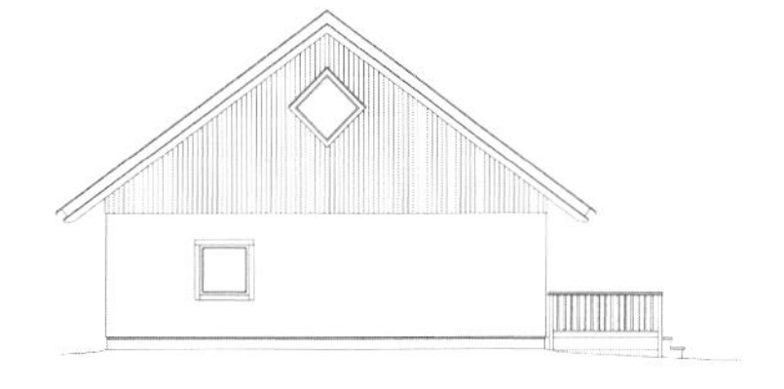

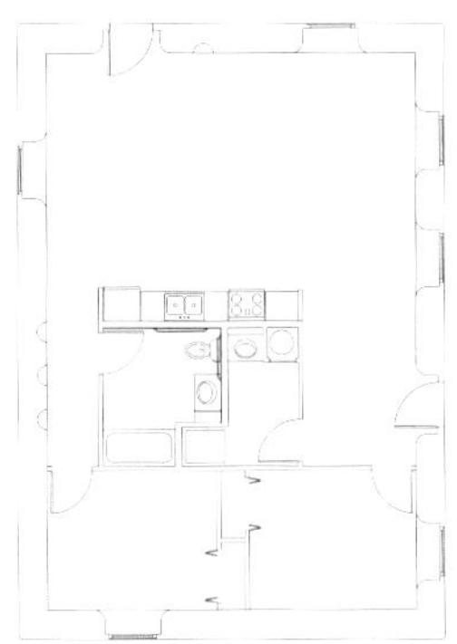

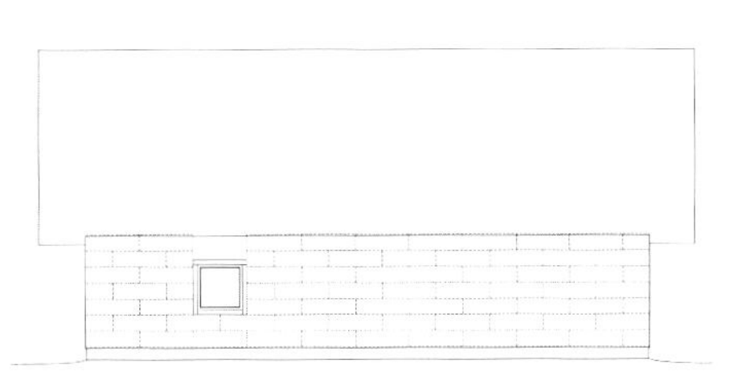

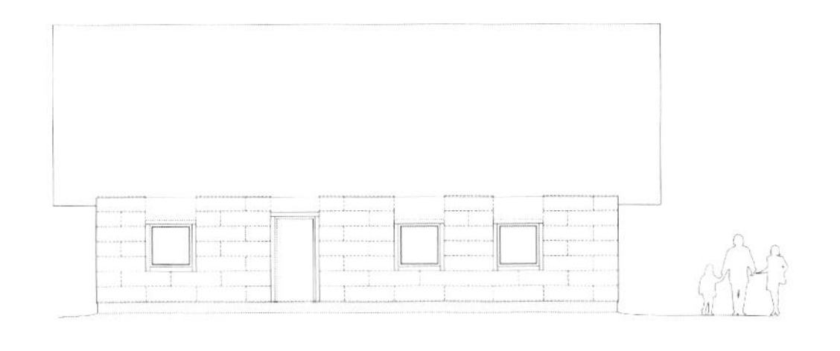

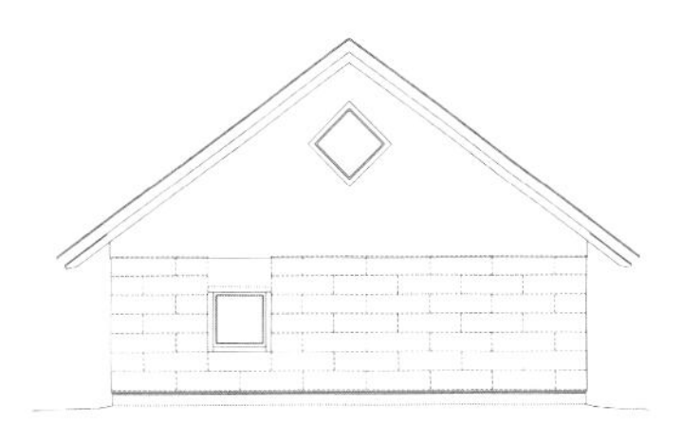

Hopi Straw Bale Home Prototype

Through a participatory design process with a Hopi tribal elder and her family, the basic straw bale home shown as Plan 4 has been geared to a specific family's needs and to the appearance of nearby historic dwellings. This two bedroom/one-and-a-half bathroom home contains a kitchen/bath core featuring a laundry room with an extra sink to ease the morning rush. The low-slope shed roof, the window and stucco treatments, and various finish details respond to the local vernacular and shelter approximately 1,100 square feet of living space

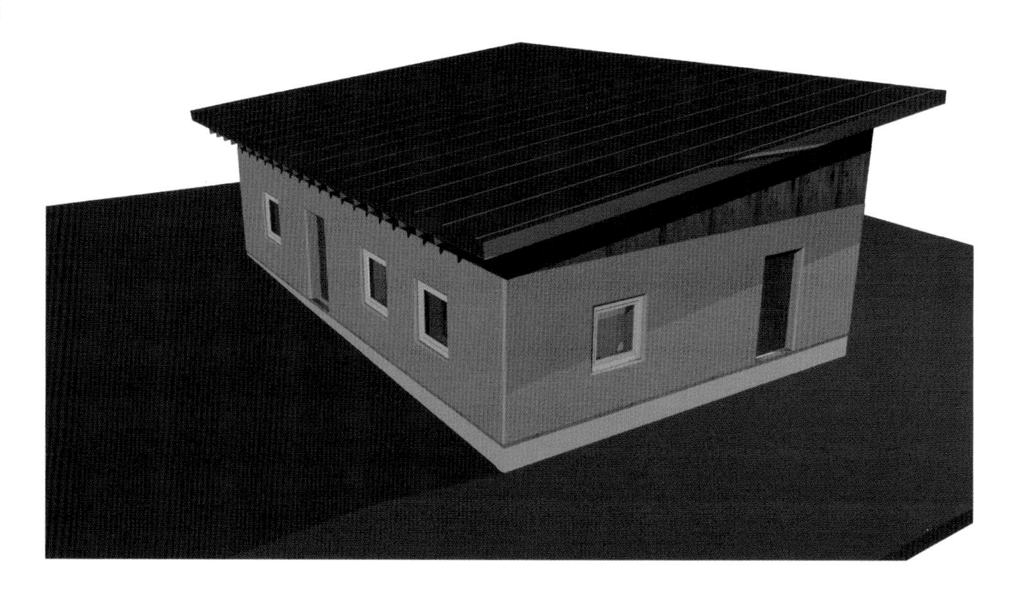

Straw Bale Navajo Hogan Prototype

Through conversations with Navajo tribal elders, the traditional Navajo dwelling—the *hogan*—is interpreted here as a load-bearing straw bale home. This straw bale Hogan consists of a kitchen, a bathroom, and associated living areas within an open plan of approximately 1,100 square feet. At this scale, the building would have a similar cost and materials list as the rectangular prototype house shown on the previous pages while responding to longstanding Navajo dwelling principles such as an east-facing entry and a central hearth and ceremonial area.

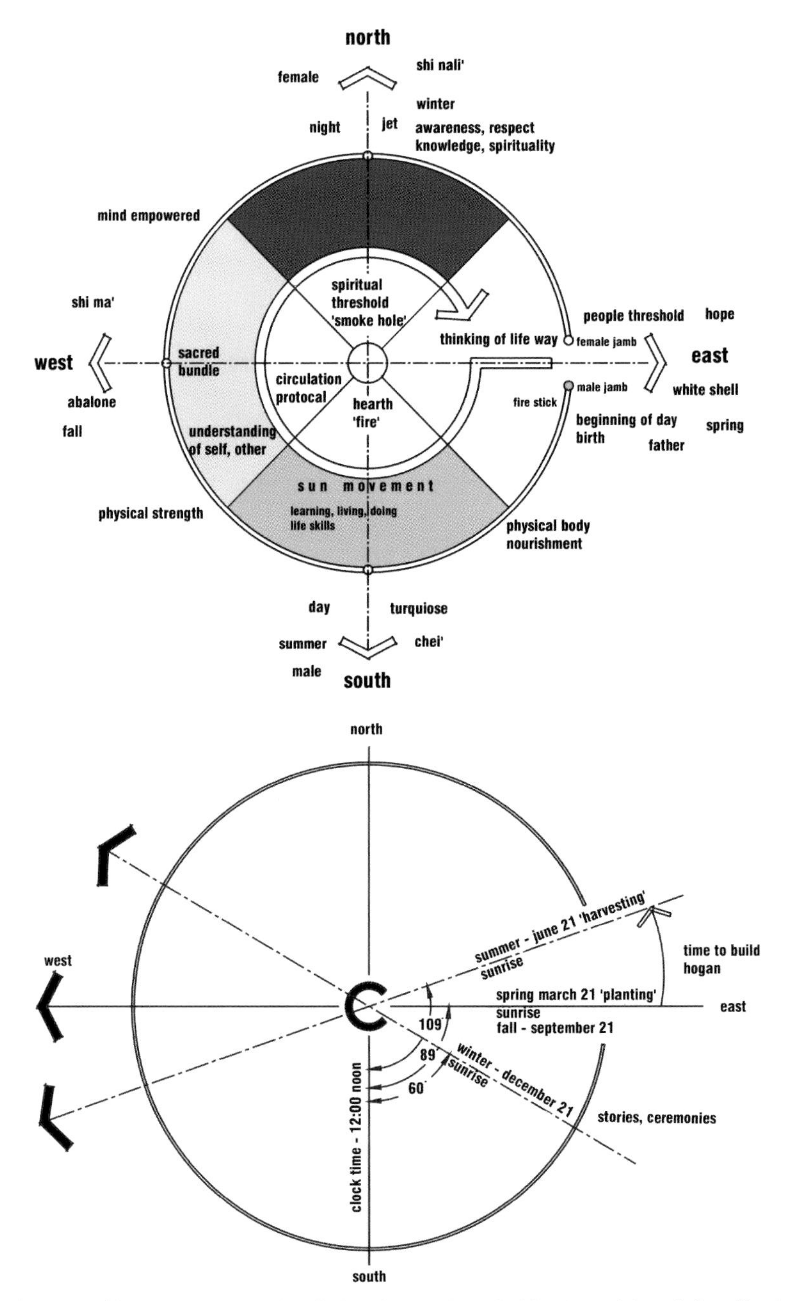

Top: A diagram shows some of the correspondences traditionally drawn between the cardinal directions and the symbolic and functional regions of the Navajo hogan.
Bottom: A diagram suggests the traditional relationship between the agricultural season and the preferred period for Navajo hogan construction.

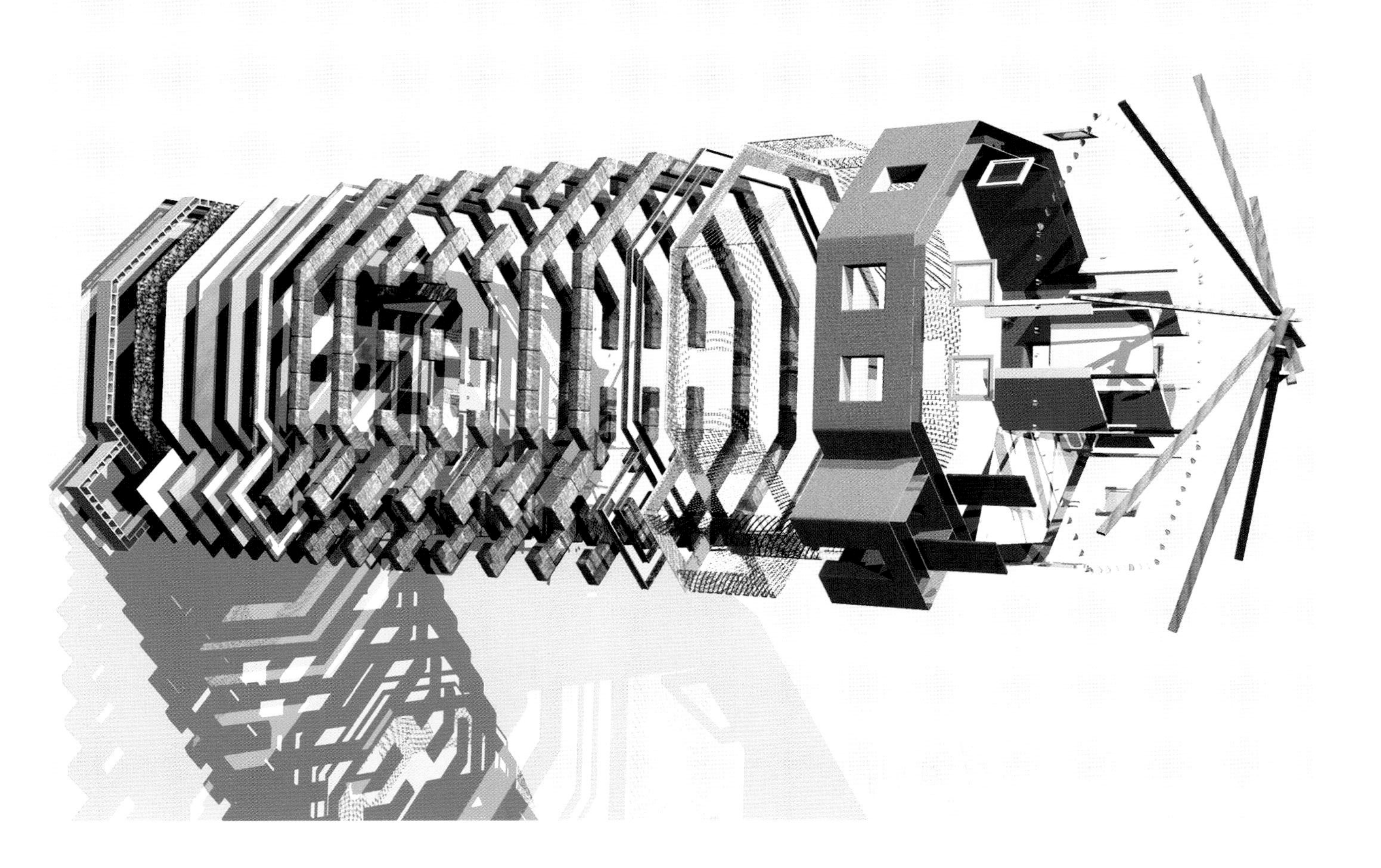

APPENDIX C | **TIME AND LABOR**

Red Feather projects typically enjoy the good fortune of having about forty community participants and volunteers on a site at any given time. One problem is how to keep everyone busy and in a position to learn each phase of the straw bale construction process. As an owner/builder, however, you may not have access to this many helpers. In such a case, you must be prepared to get your straw bale walls quickly covered in case inclement weather arrives and threatens the quality and longevity of your building.

Below is a typical Red Feather schedule for building a 1,000–1,200 square foot straw bale building in about three weeks' time with approximately forty people with various levels of construction experience. If you are working with a smaller crew, you can get a good idea of how to phase your work from this schedule. You can also extrapolate from the days required for these tasks to adjust your schedule to the number of workers available.

In practice, we often lengthen this schedule in order to have more time for the educational and cultural programming we consider as important as the work itself. Also we are careful to give a full day off whenever we see that the work team is fatigued or in need of rest. No matter how much you want to see a project completed, bear in mind that when people are tired, mistakes and accidents happen. By extending the building schedule by even a few days, workers will enjoy a safer job-site environment, and will learn more about construction techniques.

Schedule

Note: the following schedule begins with a finished foundation already in place.

Day 1 Snap lines for Roof Bearing Assembly (RBA). Build RBA sections as well as door and window bucks.

Day 2 Install base plates on slab. Place insulation between base plates. Set door bucks. Build ramp and scaffolding, and start bale wall.

Day 3 Stack bale walls. Set window bucks within bale walls. Note: Door bucks were installed at the time of base plate installation.

Day 4 Complete bale walls. Set RBA atop walls.

Day 5 Set trusses. Start roof framing.

Day 6 Frame roof. Begin exterior lath. Lay out plumbing and electrical locations.

Day 7 Sheet roof with plywood. Stock roofing materials on roof deck. Start drywall on ceiling.

Day 8 Start roofing. Complete ceiling drywall. Lay out interior walls. Insulate roof. Start taping ceiling drywall.

Day 9 Complete roofing. Stucco exterior—apply scratch coat. Frame interior walls. Start interior lath. Ceiling drywall first sanding and second mud coat.

Day 10 Stucco exterior—apply brown coat. Rough-in plumbing and electrical service. Continue interior lath. Finish sanding ceiling drywall. Roll primer paint coat on ceiling.

Day 11 Stucco exterior—apply finish coat. Complete rough-ins of plumbing and electrical service. Inspect and complete all lath work.

Day 12 Stucco interior—apply scratch coat to bale wall interior. Hang drywall on other interior walls. Start exterior trim.

Day 13 Stucco interior—apply brown coat. Complete interior drywall. Begin drywall taping. Continue exterior trim work.

Day 14 Stucco interior—apply color coat. Complete drywall first sanding and second mud coat.

Day 15 Finish sanding interior drywall. Begin interior painting. Complete exterior trim.

Day 16 Begin interior trim. Set cabinets and doors. Begin electrical trim.

Day 17 Continue interior trim and detailing. Set plumbing fixtures. Complete electrical trim.

Day 18 Complete plumbing trim. Install hardware. Clean up and punch list of remaining tasks.

Good luck with your building!

APPENDIX D | **TOOLS AND MATERIALS**

The following list represents the general material requirements for the typical phases of a building project similar to the one presented in this book. This list is not a complete listing of materials, since this book is intended to illustrate a building type which could vary significantly in size and scope. Also, on any given project, different materials will be substituted, salvaged, or omitted by the owner/builder or contractor. This list is intended to provide a point of departure as you produce early materials schedules for a straw bale building project. The owner/builder or contractor can use this list as a reference while preparing materials take-offs once the building size has been determined. We have tried to list some of the tools that are particularly appropriate to the straw bale process and some of those that are not typically found in the carpenter's toolbox. This is not, however, a comprehensive tools listing.

Tools and Materials List

The list below is broken down into phases that relate to the numbered chapters of this book. For each phase a percentage is listed to give readers a sense of the amount of work involved relative to the entire project. For example the foundation work of phase one represents approximately fifteen percent of the entire project work.

Chapter 1: Foundation

(15% of total project work)

- Forming material, concrete, insulation, reinforcing steel, and other foundation materials per the recommendations of project, soils, and structural engineers.

Chapter 2: Roof Bearing Assembly

(8% of total project work)

- I-beam floor joists (TJIs) for RBA sides and intermediate blocking
- ½″ plywood sheathing for RBA top and bottom planes
- Deck adhesive and glue gun
- 16d and 8d vinyl coated framing nails, or if available pneumatic nails with nail gun, air compressor, and hoses
- Galvanized coil strapping for attachment between base plates and RBA (to be installed after roof is completed)

Chapter 3: Base Plates

(5% of total project work)

- 4x6 pressure-treated exterior plate stock
- 4x4 pressure-treated interior plate stock
- 2x4 pressure-treated material for inter-plate channel
- 1x4 pressure-treated material for inter-plate channel
- Deck adhesive
- Fasteners and fastening tools for connection to slab (i.e. threaded wedge anchors, and roto-hammer with bits)
- Framing nails and traction nails

Chapter 4: Window and Door Bucks

(3% of total project work)

- 2x4 plates for window and door frames
- Plywood sheathing for buck skirts
- Deck adhesive

- 16d and 8d framing nails, or pneumatic nails, gun, etc.
- 1½″ wood or drywall screws for corners of buck skirts

Chapter 5: The Straw Bale Wall

(20% of total project work)

- 3-string straw bales
- Rebar cutter/bender
- #4 rebar for wall pins
- #3 rebar for forming corner bale staples
- Short handle sledge hammer for driving rebar pins
- Some type of home fashioned bale hammer (see photograph in Chapter 5)
- Buckles and nylon strapping for custom bales
- 2x10 and 2x4 lumber for corner braces/wall guides
- 6′ level/straight edge

Chapter 6: Above the Bale Wall

(20% of total project work)

- Roof trusses
- Vented bird blocks for soffit ventilation
- 2x4 lumber for bargeboard outlooks and truss bracing
- Hurricane truss clips (for example, Simpson H-1 truss clips)
- Cedar fascia stock for fascia trim and bargeboards
- 1x2 or 1x3 cedar for rake trim on bargeboards
- ½″ ACX plywood for exposed overhang areas
- ½″ plywood or OSB sheathing for roof planes
- H clips for horizontal seams in plywood/OSB sheathing
- 16d framing nails for trusses and truss bracing, 8d framing nails for plywood, or 8d pneumatic equivalent if available
- Slap stapler and staples
- Ridge venting assemblies/materials
- 1x2 metal drip edge (optional at roofing contractor's discretion)
- Roofing underlayment (for example, ice and water shield)
- Roofing materials (metal) and associated fasteners
- Metal flashing for valleys and skylights as dictated by design
- 5/8″ type-X GWB gypsum drywall for ceiling
- R-38 insulation for attic area

Chapter 7: Lath

(7% of total project work)

- 2.5 galvanized metal lath
- 1 ½"x17 gauge stucco netting (galvanized chicken wire)
- ¾" J-channel with weep screed
- ¾" J-channel
- Tie wire
- Landscaping pins
- House wrap to wrap window and door bucks
- 1x4 lumber stock, or plywood scraps for alternate to J-channel for interior base screed

Chapter 8: Stucco

(15% of total project work)

- Portland Cement as indicated by recipe (with fly ash if possible)
- White Portland Cement as indicated by recipe
- Polypropylene fiber for reinforcing stucco mixture (optional)
- Lime as indicated by recipe
- Sand as indicated by recipe
- Coloring pigments for stucco or paint (optional to personal preference)
- Gasoline powered stucco mixer (and gas for operation)
- Supply of clean pressurized water with hose and sprayer nozzle to mix stucco, slake walls, and clean tools
- Hawks, trowels, scratch coat comb, and sponge trowels
- Gloves and filtration masks
- Square ended shovels
- Rubber mallet
- 5 gallon buckets in sufficient numbers to keep the process flowing
- A pre-measured and marked container for adding lime to the mix
- Hose of sufficient length to stretch from water source to mixing station
- Several wheelbarrows

Chapter 9: Interior Walls

(5% of total project work)

- 2x4 plate stock
- 2x4 studs
- Pneumatic gun, hammers, and nails
- Drywall
- Water resistant drywall for bath
- Drywall tape and joint compound
- Drywall screws and nails
- Mud trays and taping knives

Chapter 10: Finish Details

(2% of total project work)

- Lumber and fasteners for porches, decks, accessible ramps, or other features
- Exterior and interior wood trim paints and/or stains
- Latex caulking, sealants, backer rods
- Interior flooring materials (if used)
- Electrical and plumbing materials
- Plumbing fixtures
- Appliances and cabinets
- Window and door assemblies and hardware
- Window and door trim
- Baseboard and/or cornice trim
- Exterior redwood or cedar trim
- Head flashing for above windows
- Sealant (for example NP-1 sealant or similar)
- Moisture barrier (for example Moistop or similar)

ICBO ES ER 5582 UPC 180°/100 psi 200°/80 psi LM40799 17/02/0
LOT# 2323 LINE#4 720feet

APPENDIX E | **RADIANT FLOOR SYSTEM**

Courtesy of Art Fust

First, it should be noted that radiant heating systems designed for straw bale buildings require fewer materials and system components because of the efficiency of the straw bale envelope. This efficiency results in less time and money spent during installation and in lower long-term heating costs.

The following is a simple, low-cost way to provide radiant heating for a straw bale home with a concrete floor. Such a system can provide the heating needs of a building in any area of the continental United States.

Most concrete slab installations will require only perimeter edge insulation or a "thermal break" from the cold foundation. The 2" thick rigid foam insulation recommended in this book will be adequate—especially if a shallow frost-free foundation is used. However, projects sited on a high water table, or on bedrock, and building designs that specify high R-value floor coverings, or low setback temperatures that require quick warm-ups will require appropriately sized under-slab insulation.

Heating Load and Pipe Spacing

You will need to calculate the basic heat loss of your house in BTU/H (British thermal units per hour) on a room-by-room basis if you want to buy only the materials necessary for an optimally efficient system.

Once the heat loss for an area is known, the following chart can be applied to determine the pipe spacing for a bare 4" concrete floor slab with a design room temperature of 65°F.

Supply Water Temp - BTUH/ft² of floor area emitted

Pipe Spacing	90°F	100°F	110°F	120°F
12″	24	34	44	54
18″	20	29	37	46
24″	17	24	32	40
30″	14	20	27	34

Hot water for a straw bale residence can be supplied from a slightly oversized domestic water heater (your required domestic hot water load + the house heat loss that you calculate). If you use a domestic water heater as your hot water source, the temperature will usually be in the 110˚F to 120˚F range. A tempering valve can also be installed to reduce the temperature of the water circulating in the floor. The maximum allowable water temperature for a concrete floor is 150˚F.

If you divide the room heat loss by the square footage of the room, the result will be how many BTUH/ft² of floor area will be required. Then, from the table above, you can determine what your pipe spacing will be at the domestic hot water temperature you selected. Do not be alarmed if you discover that 24″ or 30″ spacings will heat your house adequately. The only downside to widely spaced pipes is that it will create larger variations in floor surface temperatures—that is, widely spaced radiant loops create hotter areas of floor above the pipes with comparably cooler areas in between.

Loop Length and Zoning

Next you will need to determine the maximum length of your piping loop. Using ½″ diameter piping, about 333 feet is the maximum possible loop length; a maximum loop flow rate of 1.2 gallons per minute (GPM) is suitable in this case. A 333 foot loop at a 1.2 GPM flow rate will require about 13 feet of head, or pressure drop, to operate properly. If the loop is only half as long (167 feet), then the flow rate will be about half as much (.6 GPM), with a significantly lower pressure drop (2 feet). Two or three small rooms with the same BTUH/ft² heat loss can be zoned on a single continuous loop or a very large room can be heated by two or more separate loops.

All loops will need to have a minimum of one ball valve to balance the flow for the system. A ball valve works by throttling (reducing) the flow through shorter loops. This same valve can also be used to close off flow to unused areas that do not require heat at certain times.

Piping Installation

Normal installation is accomplished by placing 6" square wire mesh that will be covered by poured concrete. Tie the piping to the wire mesh grid at about 24" on center with wire or nylon cable ties. Optimum placement of the piping and mesh is in the center of a 4" deep slab pour.

Important: To prevent rupturing piping when drilling anchor holes for interior walls, cabinets, and the like, be sure to place loops only where there is open floor space above them. Between room areas piping should typically be run through doorways to avoid damage from subsequent construction. Interior partitions and other features which will be fastened to the concrete floor should be planned in advance and laid out before placement of the piping and the pouring of the floor slab to avoid danger of damaging the pipes and the extra work that accessing a rupture and splicing it would require.

Pump and Control Components

Most reasonably sized homes (less than 2,400 square feet) can be heated with one circulating pump and the appropriate balancing of the various piping loops (as described above) through the use of a piping-loop ball valve. The pump size is determined by adding up the GPMs for all of the loops and the longest loop pressure drop (13 feet if there is a 333 foot loop). For example: three loops of varying lengths with one loop at 333 feet = 2 to 3 GPM @ 13' head. The simplest control of the pump is typically by a line voltage thermostat (120 volt) located at an appropriate central location. However a 24 volt programmable thermostat and relay (to operate the pump) can be very useful for automatic set back operation.

Heating System Piping Diagram

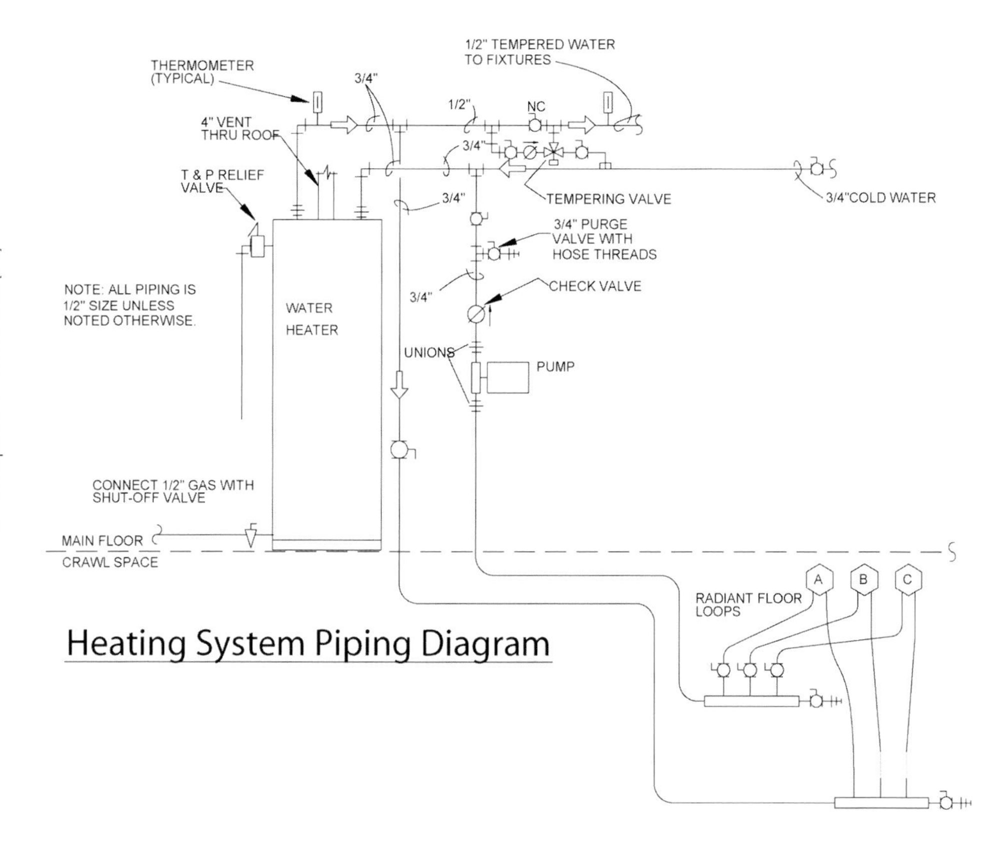

As this system utilizes potable water, piping, pumps, valves, and other components need to be non-ferrous and appropriate for potable water. Such a pump will be made of bronze or stainless steel and will be more expensive than a cast-iron one.

The application described above is very basic with simple components that work well together. Certain considerations must be made, however, for high heat loss areas, multiple zone installations, and differences in floor covering R-values. When such a system is optimally designed, it will cost less than any other central heating system available, and will not only be invisible, but more efficient, comfortable, and quiet.

Mark interior wall locations before placing radiant loops so that the tubing will not be punctured by plate fasteners during future framing.

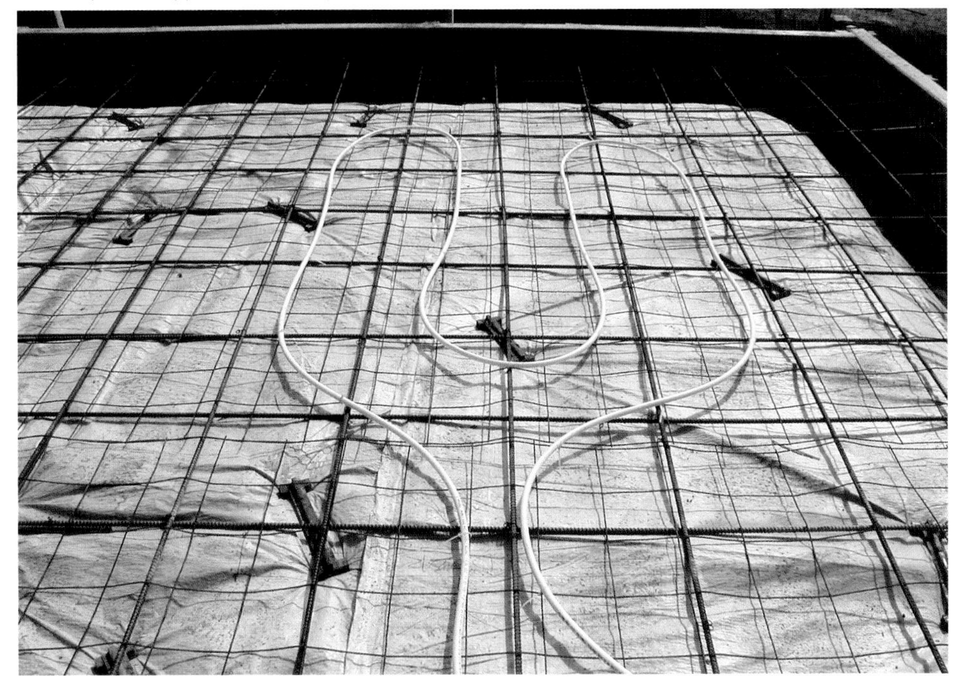

APPENDIX F | **STRAW BALE REFERENCES**

With a set of plans in one hand and this book in the other, you have an overview of the construction methods Red Feather uses to make straw bale buildings. By making this process accessible and straightforward, it is hoped that a wide range of people and communities will be able to develop skills, build houses, and feel comfortable with straw bale as a building material.

In any type of building—especially one as personal as a home—you have many choices. In order to facilitate the process of building with straw bale, this book is geared to building a house relatively quickly and inexpensively. In doing so, we have omitted many options that are possible in the realm of straw bale construction. Fortunately, there are readily available publications and resources able to remedy this deficit. With the following list of books, you can satisfy your curiosity with respect to more elaborate house forms, more options in the construction process, and more technical explanations of straw bale building methodology.

Straw Bale Construction Resources

King, Bruce. *Buildings of Earth and Straw: Structural Design for Rammed Earth and Straw Bale Architecture.* Sausalito: Ecological Design Press, 1996.

Lacinski, Paul and Michel Bergeron. *Serious Straw Bale: A Home Construction Guide for All Climates.* White River Junction: Chelsea Green Publishers, 2000.

Myhrman, Matts and S. O. MacDonald. *Build it with Bales: A Step-by-Step Guide to Straw Bale Construction.* Tuscon: Out on Bale Publishers, 1999.

Steen, Athena, Bill Steen, and David Bainbridge with David Eisenberg. *The Straw Bale House*. White River Junction: Chelsea Green Publishers, 1994.

Straw Bale Related Websites

CASBA—California Straw Building Association
www.strawbuilding.org

DCAT—The Development Center for Appropriate Technology
www.dcat.net

EBNet—Ecological Building Network
www.ecobuildnetwork.org

The Last Straw Journal
www.thelaststraw.org

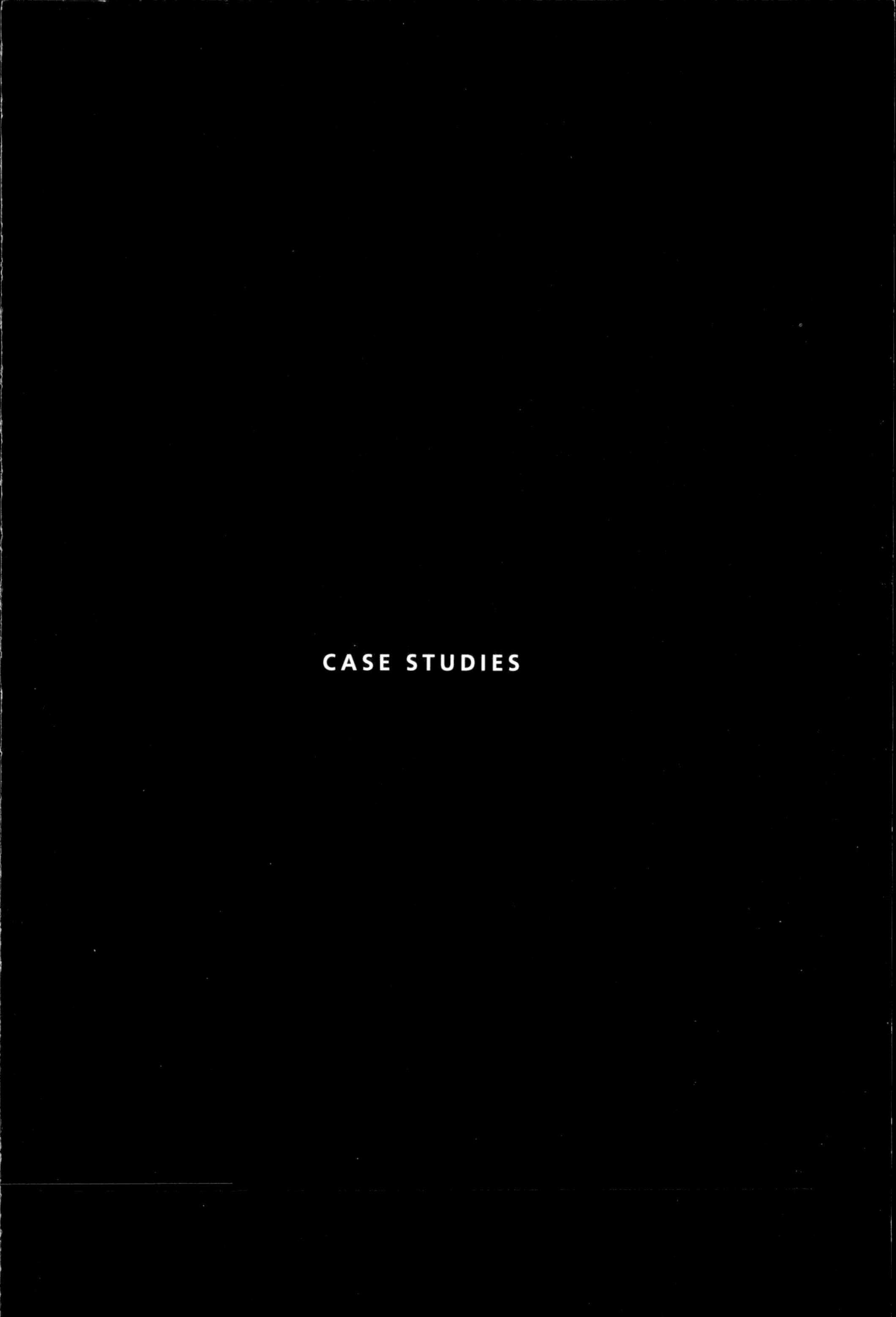

CASE STUDIES

Feather built this one for an elderly Oglala Lakota couple.

Case Studies

There are different ways to build a house, and straw bale homes are no exception. Over the past five years Red Feather has constructed a variety of buildings, each of which possesses unique qualities because of its site, materials, community interaction, or design.

The Red Feather projects illustrated here represent a range of building options. Each project suggests a way of personalizing the prototype house presented in this handbook. Taken together, these images also represent Red Feather's history as a grassroots effort in community improvement, a history that began a decade ago when Red Feather installed wheelchair ramps and repaired the homes of tribal elders, and then expanded to constructing tribal housing and community facilities.

Improving Existing Homes

Several of Red Feather's past projects involved winterizing and renovating existing homes. In terms of sustainability, a group cannot do much better than improve buildings that already exist. By finding new uses for existing structures, and by restoring buildings that have gone too long without maintenance, Red Feather inexpensively transformed substandard housing into viable homes. Because a vast number of sub-standard homes exist on reservations that can be repaired in this manner, it is possible, using Red Feather's techniques, to provide significant amounts of inexpensive and decent housing for those who need it. Moreover, straw bale technology offers excellent and far-reaching options for innovative renovation, especially in cold climates. Buildings that are inadequately insulated, for example, even trailer homes, can acquire new exterior insulating layers of straw bale and stucco that drastically reduce suffering, and heating bills.

The homes Red Feather builds from the ground up are similar in several ways. Each typically features a load-bearing straw bale wall system. Each is constructed by community members who work together with Red Feather volunteers. Each uses similar materials and a similar process—a "blitz-build" that lasts about three weeks on average. And each has a similar appearance. Most importantly, all are modest, yet attractive; inexpensive, but of high quality; simple, but noble.

Look at the following projects for differences. Each building follows a different plan and has a different profile. Roofs, windows, doors, and porches vary in form and configuration. Each building has a specific relationship to its site and its community. And each has interior walls flexibly designed within the load-bearing straw bale envelope. Were a community to build several straw bale structures at once, variations of these sorts would go a long way toward making each home unique, and to tailoring the home prototype presented in this handbook to the family who would build it. The final project shown—The Turtle Mountain Environmental Research Center—goes into more depth, providing a more intimate picture of a build.

Crow Nation Community Study Hall

Four Crow junior-high school students, inspired by a previous Red Feather/University of Washington straw bale project, entered a Bayer/National Science Foundation (NSF) competition. These girls won the top prize of $25,000 for their research and presentations on straw bale construction. Their NSF prize money, together with matching funds from Oprah Winfrey and the assistance of Red Feather, helped make their dream of having an efficient straw bale community study hall come true.

Colorful treatments of walls and porches enhance this building's simple design. Door and window trim set flush with the wall surfaces minimizes the chances of water infiltration.

This cabin is the home of an Oglala Lakota tribal elder whose father built it when his family was forced onto South Dakota's Pine Ridge Reservation. Red Feather renovated and winterized the cabin and added water and electrical service.

Crow Nation Community Study Hall, Montana

Northern Cheyenne Reservation, Montana

Pine Ridge Reservation, South Dakota

Northern Cheyenne Reservation

This straw bale home was financed with a home mortgage from the United States Department of Agriculture (USDA) Rural Development Program. During the course of a three-week summer build, the new homeowners worked alongside Red Feather partners and volunteers to build a two-bedroom, one-bathroom straw bale home. This project represents Red Feather's ongoing collaboration with the Northern Cheyenne community to address housing shortfalls.

As a load-bearing straw bale house sited to take advantage of the sun's warmth, the south facing facade has several windows to allow winter sun into the interior where stone finishes absorb and store solar heat. The primary entrance porch is also located to take advantage of the warmer southern exposure. A small hill and a stand of coniferous trees protect the house from cold northern winds.

Pine Ridge Reservation

Oglala tribal members, Red Feather volunteers, University of Washington staff and students, and members of the Adopt-A-Grandparent Program built this two-bedroom home for Oglala elders. The University of Washington College of Architecture and Urban Planning provided the home design, and the Department of Construction Management provided on-site project management. St. Thomas More parishioners and Red Feather Development Group supporters provided the funding.

The north side of this building has no windows, which, so to speak, turns the back of the house to extreme weather and cold winter winds streaming across the Great Plains. The attic is vented through the roof to alleviate the need for vents in the gable ends, and stepped site walls on the windward side of the house divert cold winds and drifting snow.

Crow Reservation
Red Feather built its first straw bale home—in partnership with the departments of Architecture and Construction Management of the College of Architecture and Urban Planning (CAUP) at the University of Washington—as a prototype to determine whether straw bale was a feasible way to solve tribal housing shortages. Funding was provided by Red Feather supporters as well as corporate and foundation donors.

A long site wall separates this house from an approach road and a parking area. The relatively steep pitch of the roof provides the significant living space of a loft. Gable ends are clad in a rustic board-and-batten style that works well with the salvaged posts supporting an entry porch."

Northern Cheyenne Reservation
The family members who helped to build this home discovered straw bale construction technology while visiting the neighboring Crow reservation where they toured a Red Feather/University of Washington project. The family then procured a mortgage for materials through the USDA Rural Development Program, and Red Feather, the University of Washington, and Penn State University provided the volunteer team, students, and technical support that helped them build this four-bedroom, two-bathroom house. This, Red Feather's largest residential project to date, is also the first straw bale home built in the Northern Cheyenne Nation. It is also the first straw bale home to be financed by the USDA.

A front porch serves as an exterior "room" in temperate weather. Windows are inset slightly on the exterior side to provide attractive detail, but this feature must be carefully built and maintained to avoid water infiltration in extreme climates. The metal roof, attractive and long lasting, also permits rainwater collection for use in plant cultivation.

Crow Reservation, Montana

Northern Cheyenne Reservation, Montana

Northern Cheyenne Literacy Center, Montana

Turtle Mountain Envrionmental Research Center, North Dakota

Northern Cheyenne Literacy Center

This community facility was a collaborative effort between the Northern Cheyenne's Chief Dull Knife College, the University of Washington, and Penn State University, with partial funding and equipment support from Red Feather Development Group. The resulting literacy center offers the power of reading to tribal members as well as providing an excellent example of volunteer-friendly straw bale construction.

Straw bale site walls tie this center to its site. Stepped site walls with integrated benches also frame an access ramp and create a gathering space in front. A warm integrated color treatment gives the stucco warmth, and a sculptural entry, adding interest to the facade, protects the interior during the winter.

Turtle Mountain Environmental Research Center

Red Feather's most ambitious project to date was the result of the hard work of Red Feather volunteers, the people of the Turtle Mountain Reservation, and Turtle Mountain Community College. Financed by a USDA Community Facilities Grant and a grant for a solar photovoltaic power system from the North Dakota Department of Commerce, the building houses classroom space adjacent to the tribal college's nature trail system, woodlot, and agricultural fields.

The building has a strong passive solar design and is sheltered by a birch-studded hill to the north. The center's materials are a ready text of sustainable design: a high-volume-fly-ash concrete floor with radiant heating, a load-bearing straw-bale wall, additional insulation of post-consumer cellulose, a roof composed of structural insulated panels (SIPS), and interior partitions of compressed sunflower seed hull panels—an agricultural byproduct material from North Dakota.

Turtle Mountain: The Anatomy of a Build

Turtle Mountain Community College (TMCC)
Straw Bale Environmental Research Center (ERC)
Turtle Mountain Reservation, North Dakota, 2004

When Red Feather constructs a house or a community facility, it creates more than just a building. A new group of caring, hardworking, and spirited individuals comes together, reaching beyond themselves to provide shelter and cross-cultural understanding in an area of need. Red Feather educates community members and volunteers, and in sharing the local culture's songs, stories, and food, they form new friendships. Building the Turtle Mountain Community College (TMCC) Environmental Research Center (ERC) during the summer of 2004 provided an excellent example of how friendships, awareness, knowledge, and understanding were built alongside an important straw bale tribal college facility.

The Builders:
Collaboration with TMCC's staff, students, and its president, Gerald Monette, guaranteed some local community involvement in this project. But the residents' participation greatly expanded throughout the build as a result of visits from U.S. Senator Byron Dorgan and several hundred tribal members, including three tribal council members, novelist Louise Erdrich, a local drum group, and local craftspeople, artisans, and teachers, who offered hospitality, hands-on skills, and construction experience. In addition to the hard work of many tribal members, others welcomed the work party with delicious home-cooked soups, vegetarian dishes, bannock and bangs (fried bread), and bison meat. Each day was rich in opportunities for tribal members and volunteers to get to know one another while sharing in the construction process.

The TMCC build also featured a new and important approach to community involvement in the form of a grant that funded the full-time participation of four of TMCC's Construction Trades department students, Jeff Grant, Jacob Laducer, David LeDoux, and Mike Martin, along with their instructor, Luke Baker. They brought considerable knowledge to the site, facilitated the involvement of other students and tribal members, served as ambassadors between Red Feather volunteers and the people of Turtle Mountain, and added straw bale construction and sustainable building knowledge to their building repertoires.

Local press coverage helped draw North Dakotan volunteers and visitors from places as distant as Grand Forks and Minot. News teams from Minot television stations KXMC-13 and KMOT-10 documented the build and broadcasted it throughout the region. Local newspapers such as the Grand Forks Herald, Minot Daily News, Turtle Mountain Times, and Turtle Mountain Star also covered the build. As a result, many visitors came by on weekends to experience a straw bale building and to ask questions. Red Feather staff and volunteers also provided tours for numerous curious site visitors.

As diverse as the community involvement was the varied backgrounds of the seventy Red Feather volunteers who worked on the ERC in July 2004. Red Feather hosted volunteers from four countries, sixteen states, and at least six other tribal nations, including Spirit Lake (North Dakota), Standing Rock (North and South Dakota), Pine Ridge (South Dakota), Cree (Canada and North Dakota), Crow (Montana), and Assiniboine/Gros Ventre (Montana).

The Build:
The participants lived, worked, and ate with one another for periods of one to three weeks. Camp life meant sleeping in tents, bathing under open-to-the-sky solar showers, and taking turns cooking in the main tent. Although there was a great deal of work to be done, volunteers and community members also found time at meals and at day's end to get to know one another, play music around the campfire, and pursue various recreational, cultural, and educational

activities. The Turtle Mountain volunteers had access to a tribal college campus and its facilities, as well as a large lake with canoes, hiking trails, beautiful terrain, diverse flora, and, among other fauna, two bison herds.

The build offered educational opportunities: the construction process—the heart of the build—served as an open "classroom" where one could acquire construction-related skills, a deeper understanding of sustainable building techniques, and general confidence boosting. Straw bale educators Matts Myhrman and Judy Knox joined forces with Red Feather's Mike Kelly and Nathaniel Corum to present a broad range of issues in straw bale construction. And though the central focus of this Red Feather build was straw bale, participants also learned a great deal about slab foundations, radiant heating systems, sustainable building materials, and conventional carpentry.

Additional educational opportunities were organized when time permitted. One evening was graced with the music and storytelling of Ed "King" Johnson—a Métis fiddler. On several occasions, those at the build had the chance to walk through the North Dakota woods with Ojibway elder Marvin Bald Eagle Youngman and share his knowledge of native plants and their traditional medicinal uses. Scott Frasier and Teresa Cohn of the Bozeman-based nonprofit Native Waters gave lakeside presentations focused on local ecology and hydrology and the importance of water quality.

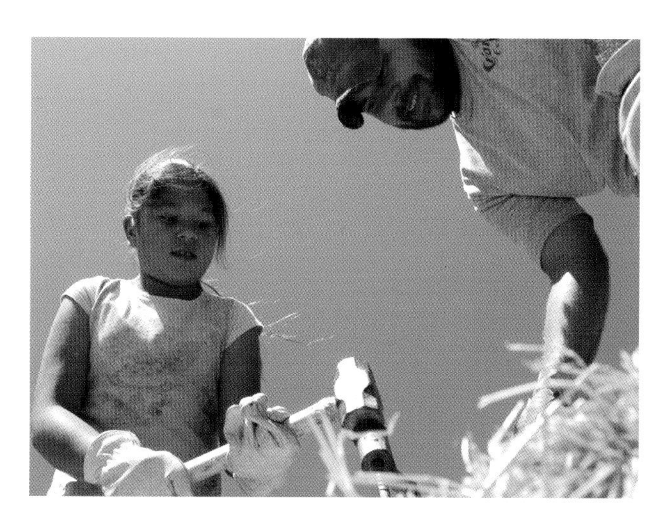

The Building:

Despite significant material and weather delays, Red Feather staff, volunteers, and local community members erected the foundation, walls, trusses, and roof of the ERC building within a three-week period. This was the first project where Red Feather staff and volunteers worked directly on the formwork and placement of the foundation and the associated radiant heating system. With the help of Matts Myhrman and Judy Knox, Red Feather staff assembled teams to raise the bale wall. Each team, led by wall co-captains, a volunteer with previous build experience, and a construction trades student from TMCC, consisted of three or more volunteers or community members. The walls went up quickly but had to be protected with tarps each night because of the threat from North Dakota's wind and rain. The true spirit of teamwork became evident when large groups of workers put heavy box beams and roof trusses in place. These crews completed the roof structure and stuccoed the exterior walls before the end of the third week. Many farewells took place as camp was struck, and the new building was prepared for the onset of winter.

Red Feather Development Group

Red Feather's work is done in the spirit of people helping people. We are in this world together, and as anyone involved in service knows, the benefits of generous work accrue to all involved: to those who have a new home in which to share food, love, and life with their families, and to those who help finance, design, and build that reality. After hearing about several tribal elders who froze to death on a southwestern reservation, Robert Young founded the Red Feather Development Group in 1994. He currently serves as executive director and has led Red Feather through the successful completion of forty housing and community-based building projects on American Indian reservations in the western United States. In addition, Young embraces corporate, foundation, federal, and university partners in support of Red Feather's American Indian Sustainable Housing Initiative. Through Red Feather, he has been able to increase America's awareness of the dire conditions of reservation housing while putting buildings on the ground. Young was the recipient of the inaugural Volvo for Life Award from Volvo Cars of North America and was recently awarded the Use Your Life Award from Oprah Winfrey Angel Network.

For the past decade, Red Feather volunteers and tribal members have built new houses and community facilities, repaired existing homes, and provided elderly and disabled members with wheelchair ramps. Together, these groups have brought hope to thousands living the nightmare of poverty on our nation's reservations.

Today, with the help of Stacie Laducer, the volunteer coordinator, Michael Kelly, the construction program director, Holly Zadra, the development director, and

Nathaniel Corum, Rose Architectural Fellow and community design director, Red Feather is expanding its efforts to develop long-term, sustainable solutions for the housing problems facing American Indians.

The current situation is at once a national tragedy and a call to action. Of the two million tribal members who live on American Indian reservations, 333,000 presently have no place to call home, and thousands more live in life-threatening, substandard conditions in buildings that lack water and electricity, and where they experience the kind of overcrowding and unsanitary conditions that destroy hopes, lives, and families. Native Americans are nearly five times more likely to contract tuberculosis than non-native Americans. Diabetes, pneumonia, and influenza, likewise, are far more common on reservations. Perhaps most grim is the fact that American Indian high school students, according to a 1999 National Center for Education study, suffer the highest percentage dropout rate of all ethnic groups in the United States—in large part because many arrive at school broken by poverty and homelessness.

Current governmental and tribal housing programs create little pride of ownership, and thus homes often go without maintenance because nothing is in place to facilitate long-term change. Habitat for Humanity—which Red Feather looked to for several foundational concepts such as sweat

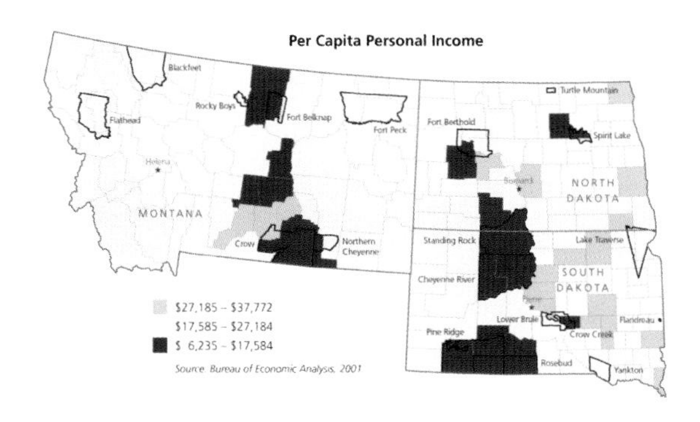

equity—recently scaled back their work on tribal lands. There is a shortage of good housing news on our nation's reservations. Red Feather seeks change—suggesting practices by which adequate housing may again become something that American Indians rely upon as they work together to build a future for themselves, their families, and their communities.

Red Feather believes that native families must take the lead in bettering their communities. Red Feather's American Indian Sustainable Housing Initiative builds on the strength of native cultural traditions and works to achieve family and community involvement in the construction of new housing and community facilities. Community members are encouraged to participate in the process at all stages. Tribal members of all ages join Red Feather staff and volunteers during construction, at mealtimes, and at the various scheduled and spontaneous educational and social events that complement a build schedule. Within the community atmosphere of the build, bridges are formed between people from all walks of life. People with divergent backgrounds find common ground and work side-by-side to create something meaningful—a new building and the many friendships that are built alongside it.

The focus of Red Feather's American Indian Sustainable Housing Initiative is to teach straw bale building techniques,

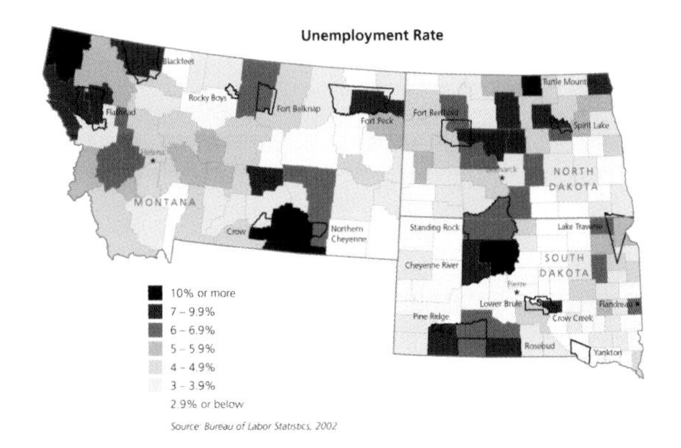

thereby establishing a framework for long-term community self sufficiency. Workshops and seminars demonstrate volunteer-friendly straw-bale construction methods. Communities build desperately needed homes and facilities. And Red Feather's partnership with major universities, tribal colleges, corporations, foundations, and thousands of concerned individuals assures that Red Feather will continue to employ the most practical, up-to-date, and sustainable construction methods available.

Such an approach allows tribal members to acquire skills and confidence that may then be passed on to the young. On recent builds, tribal members and volunteers took turns leading informational seminars, and jobs were created for tribal college students—at once empowering and educating tribal members and Red Feather volunteers alike.

Straw bale construction is a logical choice for the reservations of the western United States where the country's poorest communities face cold, dry winters and produce an abundance of wheat—and wheat straw. Sustainable and inexpensive, wheat straw, an agricultural by-product, is the basic building block of straw bale construction.

Straw bale has been used in construction since the early 1800s, and in the past twenty years, much has been done to learn the technical characteristics of straw as a building material. At present straw bale technology is working its way into building codes and into mainstream construction as people realize that straw bale buildings can be cozy,

beautiful, sustainable, and community friendly. Banks and insurers are following suit: straw bale construction has qualified for HUD, USDA, and conventional home mortgages.

Red Feather has completed a number of housing and community facilities projects in collaboration with tribal governments and United States government agencies, taking advantage of the Department of Agriculture's Rural Development Low-Income Home Mortgage program, the Indian Health Service's home sanitation and water initiatives, and the Bureau of Indian Affairs' land acquisition and ownership-verification protocols. Several major banks, including Fannie Mae and Washington Mutual, are now ready to invest in Native America.

Red Feather currently collaborates with various corporate supporters, as well as a number of universities and tribal colleges, to create affordable, sustainable, straw bale housing programs on reservations throughout the western United States. Top priority is community self-sufficiency.

Current native-based community development corporations (CDCs) play an integral role in this initiative. Participating CDCs assist tribal members with methods in home design, community planning, local volunteer recruitment, and straw bale construction. CDCs and lending institutions help native families negotiate the complex land and mortgage regulations that exist on reservations. CDCs put tribal members in positions to create positive change within their communities.

To implement effective straw bale housing programs, Red Feather identifies reservation-based CDCs and nonprofit social service programs that have the ability to incorporate sustainable building programs into their current operations. In collaboration with such groups, Red Feather's staff engages in a series of community meetings—typically throughout the winter months—to acquire tribal member input, provide straw bale education workshops, and, through participatory community design, develop tangible plans, drawings, and aspirations for new buildings directly informed by the hopes and needs of community members. Red Feather consistently reviews its building designs and construction processes vis-à-vis a given community. Future projects benefit from this because it allows CDC staff and Red Feather's design and construction directors to accommodate new methodologies and techniques while creating culturally appropriate housing.

Working directly with CDC staff, tribal members, tribal colleges, and tribal housing authorities, Red Feather helps plan and facilitate community development by utilizing tribal resources in the construction of reservation housing and community facilities. After participation in several such projects, it is possible for the community groups themselves to initiate, organize, and carry out straw bale construction projects.

The final step is follow through and maintenance. The Red Feather houses will be the first homes owned by many of the families taking residence there and their first experience with

homeownership responsibilities. Red Feather periodically checks in with homeowners and provides building materials and information so they will be prepared to tend to their homes in the years to come. In post-occupancy meetings, Red Feather queries community members, tribal authorities, and new occupants; results from these interactions are used to ensure improvements in future projects as well as in the scope of program as a whole.

At the end of the day, the measures of success are not simply the number of built structures. Although that is crucial, community involvement is also essential. Red Feather must be invited to come to work by a community. If tribal members feel they are not part of the process, community self sufficiency will not be accomplished. Care is taken to involve community members from the onset and at every level of the build process, and to secure an equitable balance between resources put forward by the organization and by the community. Only in this manner can the Red Feather buildings truly find a home in their respective communities.

To conclude, it is worth mentioning that there are several deeper layers of equity and sustainability that are important to consider if one wishes to move from building houses to making homes. Sustainability is a word that takes on new definitions daily and an important concept that must not be understood superficially. It is, at its root, a state of affairs that will allow our children's children's children the

same economic, social, cultural, ecological, and spiritual opportunities that our parents enjoyed.

Housing—having a home for one's family—is a prerequisite for most if not all sustainabilities. Given shelter, people can sustain their culture, their society, their spirituality, and their family. Home and hearth provide nourishment and a place where families may gather and children may learn languages, songs, recipes, stories and life ways from their elders—sustaining the richness of human diversity and potential. Financial equity and sweat equity need no explanation. Yet future homeowners can accrue emotional equity in the hands-on act of building a home with and for their family and community members. To make a house a home, we encourage families to imprint their homes with cultural equity during preliminary design. Spiritual equity, finally, can only reside in a home that has been both welcomed by a community and informed by a family's sense of concept, process, and creation.

CONTRIBUTORS

Corbin Plays is a Crow tribal member from Montana. He practices architectural and graphic design in New York City. His design work focuses on the future of environmentally conscious architecture. Corbin studied architecture at Cornell University.

Jonathan Corum is principal of 13pt, a precision design and development studio based in New York City. He has been a senior designer at both Font Bureau and Interactive Bureau, and design director of Online Retail Partners. Jonathan graduated from Yale College with a degree in Art and East Asian Studies.

Michael Kelly, green builder and principal of Rocky Mountain Woodwright, has been a Red Feather member since 1999 and a skilled Red Feather volunteer leader since 2002. He joined the staff at Red Feather full-time in 2004 as construction program director. Michael studied business administration at Clemson University.

Michael Rosenberg is based in Seattle and has produced portraiture, events and editorial photography for fifteen years. His background in the arts and psychology contribute to his approach—capturing what is unique about the people he photographs. Michael has a strong interest in social issues which draws him to take on projects supporting nonprofit organizations.

Richard K. Begay, Jr. is a Navajo tribal member and an intern architect currently practicing with the DLR Group in Phoenix, Arizona. His work is centered around culturally appropriate design and community service with an emphasis on native youth education. He studied architecture at the University of Arizona.

Ryan Batura comes from three generations of master carpenters and has over twenty-five years of experience in construction. His concern for native housing issues are reflected by the technical consulting, materials, and labor he has donated to charitable organizations including Red Feather—where he served as construction program director for the 2003 build season. Ryan is now principal of a construction management firm in Seattle.

Skip Baumhower works out of his photography studio in Tuscaloosa, Alabama. Each year he travels widely, living with families wherever he is working and involving himself in their communities. His goal is to document cultural diversity while creating positive change in the lives of the people he photographs.

Wayne Bastrup works as an architect in the Seattle area. He studied architecture at the University of Washington and, in recent years, has designed three straw bale structures and managed several Red Feather Development Group projects.